Fast and Fabulous!
Quick Cuisine

Fast and Fabulous!
Quick Cuisine

Judy Gilliard

THOMSON
DELMAR LEARNING

Australia Canada Mexico Singapore Spain United Kingdom United States

Fast and Fabulous – Quick Cuisine
by Judy Gilliard

Vice President, Career Education Strategic Business Unit:
Dawn Gerrain

Director of Editorial:
Sherry Gomoll

Acquisitions Editor:
Matthew Hart

Developmental Editor:
Patricia Osborn

Editorial Assistant:
Patrick B. Horn

Director of Production:
Wendy A. Troeger

Production Editor:
Matthew J. Williams

Director of Marketing:
Wendy E. Mapstone

Channel Manager:
Kristin McNary

Cover Design:
Joe Villanova

COPYRIGHT © 2006 Thomson Delmar Learning, a part of The Thomson Corporation. Thomson, the Star logo, and Delmar Learning are trademarks used herein under license.

Printed in the United States
1 2 3 4 5 XXX 09 08 07 06 05

For more information contact Delmar Learning,
5 Maxwell Drive
Clifton Park, NY 12065-2919.

Or find us on the World Wide Web at
www.delmarlearning.com or
www.hospitality-tourism.delmar.com

ALL RIGHTS RESERVED. No part of this work covered by the copyright hereon may be reproduced or used in any form or by any means—graphic, electronic, or mechanical, including photocopying, recording, taping, Web distribution or information storage and retrieval systems—without written permission of the publisher.

For permission to use material from this text or product, contact us by
Tel   (800) 730-2214
Fax   (800) 730-2215
www.thomsonrights.com

Library of Congress Cataloging-in-Publication Data

Gilliard, Judy.
   Fast and fabulous! : quick cuisine / Judy Gilliard.
       p. cm.
   ISBN 1-4180-2999-8
   1. Quick and easy cookery. I. Title.
TX833.5.G5355 2005
641.5'55--dc22

2005024220

NOTICE TO THE READER

Publisher does not warrant or guarantee any of the products described herein or perform any independent analysis in connection with any of the product information contained herein. Publisher does not assume, and expressly disclaims, any obligation to obtain and include information other than that provided to it by the manufacturer.

The reader is notified that this text is an educational tool, not a practice book. Since the law is in constant change, no rule or statement of law in this book should be relied upon for any service to any client. The reader should always refer to standard legal sources for the current rule or law. If legal advice or other expert assistance is required, the services of the appropriate professional should be sought.

The publisher makes no representation or warranties of any kind, including but not limited to, the warranties of fitness for particular purpose or merchantability, nor are any such representations implied with respect to the material set forth herein, and the publisher takes no responsibility with respect to such material. The publisher shall not be liable for any special, consequential, or exemplary damages resulting, in whole or part, from the readers' use of, or reliance upon, this material.

Donna and Mitch Baker—they saw the vision!

and

Aunt Rose . . . Aunt Rose . . . . Aunt Rose
Who's kitchen and pantry know how started it all.

Contents

Acknowledgments viii
Introduction 1
   Some Tips for an Organized Kitchen 3
   Important Kitchen Equipment 4
   Keeping Your Pantry Fresh 6
   The Basics in Shelf Life 7
   Basic Ingredients for Your Pantry 9
   Pantry Items 13
   Terms to Know 18

Condiments, Sauces, Marinades, and Spreads 21

Breads 33

Salads 39

Side Dishes 53

Main Dishes
   Salads 75
   Seafood 83
   Pasta 97
   Chicken 115
   Eggs 129
   Pork 135
   Meat 151

Desserts 173

Index 191

Acknowledgments

Donna and Mitch Baker . . . they saw the vision!

When I think about acknowledgments, I could write another book because, along the way, there are so many people that make this happen and are important to the inspiration and development of recipes.

- First and foremost, Jean Sheffield who spent hours on the phone with me going over recipes to make sure you, the reader, will understand them.
- Matt Hart, Patricia Osborn, Sherry Gomoll, and Matt Williams, to just name a few of the people from Thomson Delmar Learning who made this all come together beautifully.
- Ethan Herb and Alice Barlow with whom I've spent many hours in the kitchen.
- Mark Williams, Executive Chef from the Bourbon Grill in Louisville, who is the master of cooking with spirits.
- Andy Fotis, Food and Beverage Director of Ameristar, who shared his food "know how."
- Gary Sadlemyer for making me a part of the 1110 KFAB family.
- My friends that are family and have been there before the books began to flow! Dan Bunker, Jan and Harvey Inez, Mona and Tom Virigilo, Bill Kasal, Nathan and Lillian Shuman, Diane Lubich, Nancy and Ken Hess, Jim Wallace.
- Friends who are family: Liz, Ed, Kellyann, Kristen, Karen, and Erik Kazor. Chris, Lori and Brittany Darrington.

Acknowledgments

- Family: Aunt Rose . . . whose kitchen started it all! Dougie . . . my cousin who loves cooking and eating as much as I do; and Susan, Amanda, and Bryson Edgar.
- My sister Teri . . . whose artwork always amazes me. Cousins Chuck and Jean Gilliard, and Uncle John and Aunt Ginny Hardacre.
- Colleen Cleek and the staff at The Classy Gourmet for their wonderful support in welcoming me to Omaha.

Introduction

In this new century, the media have given much attention to what America eats, the focus being that obesity in America is at higher levels than ever before, and that obesity in children is at an all-time high. The reasons: (1) the way Americans eat, with the majority of the food consumed coming from both fast foods and convenience foods, and (2) eating on the run in a fast-paced, hectic lifestyle.

The concept of this book is to provide guidance for eating healthy, knowing how to cut calories, and eating better quality and better tasting food. With some planning, you will be able to get dinner on the table in less than 25 minutes.

The key is to keep a well-stocked pantry. Included in this book is a pantry list, with additional room to add your favorites.

Some of the recipes may take a little time the night before, such as marinating your meat overnight so it is ready for you to pop on the grill when you get home from work or, perhaps, getting ingredients ready to put in your crock pot in the morning to slow cook all day.

I hope to show you that by keeping a well-stocked pantry, you will always have a meal waiting to be made just as quickly as picking up an order from a fast-food restaurant on the way home—and far healthier and less expensive.

The new food pyramid emphasizes adding more whole grains to our diet. Whole wheat pasta has come a long way in the past five years, so if you don't think you like it, start experimenting and find ones you do like. In most recipes, I use regular pasta in the nutrition analysis. If you exchange regular pasta for whole-wheat pasta, your calories will stay about the same; however, you will increase the fiber six-fold in some cases. I think you will be surprised. When I was testing pasta dishes, friends did not notice the difference and were surprised when I told them the pasta was whole grain.

I really enjoy cooking with spirits because they add complexities and layers of flavor with ease. Have some of your favorite spirits on hand, including bourbon, vodka, vermouth and brandy.

In my testing of different oils, I found some wonderful artisan oils from California, such as those from Tutta and Bariani. Nut oils like walnut oil, almond oil, and hazelnut oil are rich in essence and add great flavor to the simplest of dishes. In contrast, grapeseed oil has a very clean taste, which is preferable for some recipes.

If you have any questions or comments please e-mail me at judygilliard@hotmail.com.

Some Tips for an Organized Kitchen

Keeping your kitchen organized and at-the-ready will enable you to come home and have dinner ready in a snap.

Salads

Buy your lettuce, bring it home, wash it, cut it up, spin it dry, and keep in an air-tight plastic bag or container. The lettuce will stay fresh for approximately five days. To prepare a salad, all you'll need to do is take out the amount of lettuce you want to use, place it in a salad bowl, and add other ingredients to it, such as:

- avocado, sunflower seeds, diced tomatoes, with rice wine vinegar
- pears, pecans, pear juice, with rice wine vinegar
- apples, walnuts, raisins, celery, with a fruit-infused vinegar
- mandarin oranges, walnuts, with walnut oil dressing
- feta cheese, diced tomatoes, with balsamic vinegar

Pasta

Keep a variety of different pastas on hand to create quick and easy pasta dinners, such as:

- prosciutto tomato pasta
- fresh basil, olive oil, shaved Parmesan cheese, and pine nuts
- capers, roasted red peppers, onions, and tomatoes

Important Kitchen Equipment

I view my kitchen equipment as investment pieces and have learned by experience that this is money well spent. It is wiser to purchase higher quality equipment, as it really makes a difference in the performance. The following equipment will make it much easier in the kitchen for you:

- **Food processor:** 12-cup capacity with wide-mouth feed tube; or 16-cup capacity food processor with a large feed tube. (This size is an excellent choice if you cook in larger amounts, as it should have a large capacity with a large feed tube, and also should have a small work bowl insert for smaller tasks.)
- **Stand mixer:** Useful for multi-tasking.
- **Indoor grill and panini maker:** I like the ones with removable grill plates, for easy clean up!
- **Mandolin:** Choose the type that will work best for you. You will wonder how you ever got along without it! This tool will enable you to create the thinnest uniform potato slices or long julienne strips that will turn an ordinary vegetable into a fancy meal!
- **Monoplane grater:** This is a simple tool that allows you to zest the skin of oranges, lemons, and limes easily without getting any of the bitter white pith. Several grates are now available. The two I use most are the **rasp** for zesting and grating fresh ginger, and the **shaver** for doing thin slices of fresh Parmesan cheese.

- **Bread baker:** This is a wonderful toy to have, as you can place your ingredients in the machine programmed so the bread is ready as you arrive home. Imagine walking in the door to the smell of homemade bread! Or, you can put a breakfast bread in the night before and wake up to fresh-made bread. I most often use my bread baker to make dough and then turn the dough into stuffed breads or pizza dough.

Keeping Your Pantry Fresh

In the past few years, I have moved quite a bit, discovering different parts of our country. This has led me to several different kitchens, from very large to very small. All this moving has forced me to keep pantry cleaning and sorting done on a regular basis. However, perhaps like you, I once lived in a home for more than 15 years and found that a once-a-year pantry cleaning was a must.

This is important for a couple of reasons. First and foremost, it keeps you organized and your kitchen fresh. Start by taking everything out of your pantry and sorting the items into categories. Most importantly, check the dates on your items to make sure they have not expired. Toss out the ones that are outdated. I am sure the number of items you have to discard will shock you. Time goes by so quickly that food can become outdated before you realize it.

For example, spices do not have a long shelf life. The best way to get the most out of your herbs and spices is to buy them whole and grind them right before serving. Spice companies are making it easier for us by putting dates on the spice jars. Buy the herbs you don't use often in small containers.

The other items to monitor are whole-grain products, such as flours and brown rice, as these can turn rancid quickly. If you put them in the refrigerator, they can last six months. Nuts also can go rancid. Keep them in your freezer to ensure their freshness. The lock-type freezer bags make it easy to take out amounts quickly and don't take up as much room as hard containers.

Once you have all the food out of your pantry or cupboard and have sorted them, clean the inside of the cupboards well. Next, restock the items on the cleaned shelves.

Do the same with your refrigerator and freezer. Put an open box of baking soda in both the refrigerator and the freezer. This will keep both areas smelling fresh.

Leftovers should be eaten within three days. Be careful to not put a steaming hot item in your refrigerator, as this will cause the temperature within the refrigerator to rise and can spoil foods quickly. Let foods cool down; then refrigerate them.

Now that you realize how good it feels to have a well-organized pantry, why wait until each spring? Plan on doing this once every three months.

I've included the Basics in Shelf Life, Pantry Ingredient List, and Pantry Items to keep your pantry stocked and to familiarize you with terms you should know. Refer to them whenever you feel the need. You may even want to take make a copy of the Pantry List Items and post it close by to help you keep your pantry stocked. This will make it easier to have fast and fabulous quick cuisine!

The Basics in Shelf Life

Baking powder	9 months
Baking soda	18 months
Butter	7 days refrigerated, 6 months frozen
Cheese, hard	4 months refrigerated, 6 months frozen
Cheese, soft	2 weeks refrigerated, 4 months frozen
Cornmeal	1 year
Eggs	according to the expiration date stamped on carton
Fish	1 day (24 hours) refrigerated, 6 months frozen
Flour, wheat	3 months, 8 months refrigerated
Flour, white	8 months

Herbs and spices	6 months, 1 year refrigerated
Honey	1 year
Mayonnaise	3 months refrigerated
Meats, ground	2 days refrigerated, 6 months frozen
Meats, whole	5 days refrigerated, 1 year frozen
Molasses	6 months refrigerated
Nuts	4 months refrigerated, 8 months frozen
Pasta, dried	2 years
Poultry	2 days refrigerated, 1 year frozen
Rice, brown	3 months, 6 months refrigerated
Rice, white	2 years
Salt	indefinitely
Shortening	8 months
Sugar, white	8 months
Sugar, brown	4 months
Vanilla	2 years
Vegetables, fresh	2 weeks
Vegetables, frozen	3 months
Vinegar	2 years
Yeast	3 months (check date on package)

Basic Ingredients for Your Pantry

- **All-purpose flour:** Finely ground flour (either bleached or unbleached) containing a moderate amount of protein; used for a wide variety of general baking and cooking.
- **Asiago cheese:** Made from cow's milk; this semi-firm Italian cheese has a rich, nutty, pungent flavor. Asiago di Taglio is aged for up to 60 days, is semi-firm and used as a table cheese. When cured for six months or more, Asiago becomes hard and is used for grating.
- **Couscous:** The separated grain of the wheat plant. When it is dried and milled, it becomes semolina flour, used primarily to make pasta. Couscous is also a North African dish made from chickpeas, various meats, vegetables, herbs, and spices. It is cooked in a large kettle with a steamer on top in which bulgur wheat is cooked as a side dish.
- **Five spice powder:** Chinese spice blend of equal parts of powdered anise, cinnamon, fennel, ginger, and clove.
- **Herbs de Provence:** A blend of dried herbs such as basil, sage, rosemary, summer savory, thyme, fennel seeds, marjoram, and lavender; traditionally associated with the Provence region of France. The herb mix is used to season meat, poultry, or vegetables.
- **Hoisin sauce:** Thick, sweet, and spicy Chinese soybean and pepper sauce, used as a table condiment and to flavor meat, poultry, and shellfish dishes; available canned or bottled in

Asian markets and many supermarkets. Transfer canned hoisin to a nonmetal container after opening. Also known as Peking sauce.

- **Mascarpone cheese:** A soft Italian cheese made from cow's milk, similar to ricotta. It can be a double or a triple cream cheese with a delicate ivory-colored, creamy consistency. Mascarpone is an important ingredient in tiramisu, and is also served plain with fruit.
- **Oat flour:** Groats that have been ground into a fine powder; often an ingredient in ready-to-eat breakfast cereals. Oat flour can be used in baking; however, it contains no gluten and must be combined with gluten flours for baked goods that need to rise.
- **Orzo:** Small pasta shaped like barley; used in soups and served as a substitute for rice.
- **Parmesan cheese:** A hard Italian cheese made from cow's milk, aged at least 14 months and as long as 4 years. It has a hard, pale-golden rind and a creamy yellow interior. The taste depends on the amount of aging, but is usually fairly sharp with a granular consistency. Parmesan cheeses are made in many countries, including Argentina, Australia, and the United States. However, purists say that none compare with Parmigiano-Reggiano from Italy.
- **Pine nuts:** Also known as pignoli or pinyons, these blanched pinecone seeds are used to give foods a rustic, aromatic flavor. They are high in fat and should be refrigerated or frozen to prevent rancidity. They can be used in sweet or savory dishes and are probably best known as an ingredient of pesto.
- **Port:** A sweet, fortified Portuguese wine, so named because it was originally shipped from the city of Oporto. Port is usually dark red, but a white port is also produced. Vintage ports,

made from grapes of a single vintage, are bottled within two years. Some of the best vintage ports can be aged 50 years or more. Port is often served as a dessert wine.

- **Prosciutto:** Salted Italian ham, aged 9 to 14 months, used as an antipasto or as an ingredient to add flavor to pasta and other dishes. Prosciutto has a dark red color, and when served as an antipasto, is usually sliced very thin.
- **Puff pastry:** Rich, flaky, multilayered pastry made by placing pats of chilled butter between layers of pastry dough, rolling out the dough, and continuing to fold and roll the dough until many thin layers of butter and dough are created. As the dough bakes, the butter releases steam, which causes the dough to puff and separate into many delicate layers; used for making croissants, Napoleons, and a variety of other creations. Puff pastry sheets can also be purchased in the freezer section of your local grocery store.
- **Rice vinegar:** Mild Chinese and Japanese vinegar made from fermented rice. There are three types of Chinese vinegar: white, which is used primarily in sweet-and-sour dishes; red, a suitable accompaniment for crab; and black, which is most often used as a table condiment. Japanese rice vinegars are colorless, mellow, and almost sweet in comparison to the Chinese varieties, and are used in preparations such as sushi. Rice vinegar is available in Asian markets and many supermarkets.
- **Ricotta cheese:** Bland white Italian cheese made famous as an ingredient in lasagna, ravioli, cannelloni, and various sweet desserts. It is slightly grainy, but smoother than cottage cheese, with a slightly sweet flavor.
- **Self-rising flour:** All-purpose flour with salt and baking powder added. One cup of self-rising flour contains $1\frac{1}{2}$ teaspoons baking powder and $\frac{1}{2}$ teaspoon salt. It can be substituted for

all-purpose flour in yeast breads by omitting the salt, and in quick breads by omitting both baking powder and salt.

- **Whole wheat flour:** A coarse flour containing the bran, germ, and endosperm of the wheat kernel, which gives the flour a higher fiber, nutrition, and fat content than all-purpose flour. Due to the fat content, this flour should be refrigerated to prevent rancidity. Whole wheat flour is not as rich in gluten as all-purpose flour, so it produces a heavier and denser baked product.

Pantry Items

Salads

Avocado

Carrots

Celery

Cucumber (English)

Herbs (fresh):
- Basil
- Cilantro
- Marjoram
- Mint
- Parsley
- Rosemary
- Tarragon

Lettuce (variety)

Radishes

Red cabbage

Red onions

Scallions

Tomatoes

Fruits

Apples

Bananas

Berries

Grapes

Lemons

Oranges

Pears

Basic Vegetables

Artichokes

Asparagus

Brussels sprouts

Carrots

Cauliflower

Corn

Eggplant

Garlic

Green beans

Leeks

Mushrooms

Parsley

Peppers (red, yellow, green)

Potatoes (Yukon Gold, red, russet, white)

Salsa

Shallots

Spinach (frozen and thawed)

Squash

Frozen

Berries (mixed)

Corn

Egg Beaters®

Green beans

Milk (low or non fat ice milk)

Peaches

Strawberries

Vegetables (mixed vegetables and stew vegetables)

Meat, Poultry, and Seafood

Bratwurst

Chicken (breasts, thighs, and legs)

Ground beef (lean)

Halibut

Italian sausage (ground)

Pork tenderloin

Salmon

Shrimp

Sole

Steak (flank and tenderloin)

Turkey (ground)

Dairy

Bread (French bread in a can)

Cheddar cheese

Cottage cheese (non fat)

Cream cheese (non fat)

Crumbled blue cheese

Milk (1%)

Parmesan cheese (fresh)

Ricotta cheese (part skim milk)

Sour cream (light)

Yogurt (non fat plain)

Wine and Liquor

Champagne

Club soda

Mineral water

Red wine

Vermouth (both dry and sweet)

Vodka

White wine

Whiskey

Whiskey Liqueur

Pasta, Rice, and Flour

Flours:
- Bisquick® baking mix
- Bread
- Pastry
- Rolled oats
- Self-rising
- Whole wheat

Pasta (various sizes and shapes)

Rice:
- Arborio
- Brown (long and short grain)
- White (long and short grain)
- Wild

Cereals

Low-fat granola

Spices, Seasonings, and Condiments

Almond extract

Arrowroot

Black pepper

Bouillon paste

Cayenne pepper

Chili flakes (red)

Cinnamon

Cornstarch

Ginger

Italian seasoning

Ketchup

Mayonnaise

Molasses

Mustard (powered)

Mustard (prepared, Dijon)

Nutmeg (whole with grinder)

Nuts (almonds, pecans, walnuts)

Oil (olive, olive spray, almond, grapeseed, walnut)

Paprika

Pepper (black and white)

Salt (sea and kosher)

Splenda®

Sugar (brown)

Sugar (white)

Sunflower seeds

Vanilla extract

Vinegar (balsamic, seasoned rice wine, variety of flavors)

Terms to Know

- **Al dente:** An Italian term meaning "to the tooth," describing the consistency of pasta cooked to a tender firmness. Usually considered to be pasta cooked just right.
- **Deglaze:** After food has been sautéed and the excess fat removed from the pan, deglazing is accomplished by adding a small amount of liquid to the pan and stirring to loosen the brown caramelized bits of food from the bottom. The resulting mixture can be used to create a gravy or pan sauce.
- **Dutch oven:** A large pot or kettle, typically made of cast iron with a tight-fitting lid; can be used in the oven or on the stovetop for moist-cooking methods such as stewing or braising.
- **En papillote:** Refers to food (usually seafood) baked inside a wrapping of greased parchment paper. As the food bakes and lets off steam, the parchment puffs up into a dome shape. At the table, the paper is slit and peeled back to release steam and reveal the food.

    This French method for cooking in paper not only saves on clean up, but is also an attractive and classical presentation technique. Start by folding a length of parchment paper to form a square; then cut along one side to form a heart shape. Oil both sides of the paper, add your ingredients, and crimp along the open edge from top to bottom.

    Place the sealed bags in a hot oven and remove once the paper has puffed up and become brown on top and around the edges.

- **Mandolin:** A hand-operated slicer with a variety of adjustable blades for thick to thin slicing; has a rectangular body with folding legs at one end that elevate it to a 45° angle. The food, usually held in a metal-guided carriage to protect the fingers, is pressed and passed against the blade to obtain precise, uniform slices or shapes.
- **Quiche pan:** Porcelain baking dish with fluted sides used for baking quiche. The dish is generally $1\frac{1}{2}$-inches deep and can be from 5 to 12 inches in diameter.
- **R-T-C:** Referred to as ready-to-cook by nutritionists. It is the raw weight of a food with the nutritional value adjusted to account for cooking losses or gains due to deboning, water absorption, evaporation, drippings, and so on.
- **Reduce:** To heat a liquid, uncovered, until the volume is decreased, the liquid thickened, and the flavor concentrated.
- **Sauté:** To cook briefly over high heat with a small amount of oil.
- **Simmer:** To keep a liquid heated just below its boiling point, so that bubbles form only at the edges of the pan.
- **Stir fry:** A Chinese method of cooking food by chopping the ingredients into small pieces and then frying them quickly in a small amount of hot oil in a very hot wok or skillet, constantly stirring and turning the ingredients.

Condiments, Sauces, Marinades, and Spreads

Barbecue Sauce à la Jack Daniel's®

Granola

Guacamole

Mayonnaise

Olive Cheddar Spread

Raspberry Sauce

Satay Dipping Sauce

Simple Tomato Sauce

Walnut Oil Mayonnaise

Yogurt Cheese

Yogurt Cream

Barbecue Sauce à la Jack Daniel's®

Serves 10

1 cup whiskey (for example, Jack Daniel's®)

1 cup ketchup

1 cup cider vinegar

1 cup brown sugar

¼ cup onion flakes

1 tablespoon red chili flakes

1. Combine all ingredients in a large saucepan.
2. Bring to a boil and simmer for about 30 minutes, or until slightly thickened, stirring occasionally.

NOTE: Makes about 2½ cups.

Nutritional Analysis (per serving, excluding unknown items)

141	Calories
Trace	Fat
0.8%	Calories from Fat
1 gm.	Protein
24 gm.	Carbohydrate
Trace	Dietary Fiber
0 mg.	Cholesterol
291 mg.	Sodium

Exchanges

1½	Other Carbohydrates

Granola

Serves 8

½ cup sugar substitute (or to taste)
¼ cup molasses, blackstrap
¾ tablespoon vanilla extract
½ teaspoon sea salt
4 cups rolled oats
1 cup almonds
½ cup water

1. Preheat oven to 275°F.
2. Line a cookie sheet with parchment paper and set aside.
3. Combine sugar substitute, molasses, and water in a 4-cup microwave-proof glass measuring cup or bowl.
4. Place in microwave on high for 20 seconds. Remove from microwave.
5. Add vanilla extract and salt.
6. Pour into a large mixing bowl.
7. Add oats and nuts. Stir until thoroughly mixed.
8. Spread the granola onto cookie sheet and bake 45 minutes to 1 hour or until golden and crunchy.
9. Cool granola completely. Store in an airtight container.

*Nutritional Analysis
(per serving,
excluding unknown items)*

312	Calories
12 gm.	Fat
33.5%	Calories from Fat
10 gm.	Protein
43 gm.	Carbohydrate
6 gm.	Dietary Fiber
0 mg.	Cholesterol
151 mg.	Sodium

Exchanges

2	Grain (Starch)
½	Lean Meat
2	Fat
1	Other Carbohydrates

Guacamole

Serves 6

2 cups nonfat cottage cheese

2 avocados

2 tablespoons salsa

2 tablespoons fresh cilantro

1. Drain cottage cheese in a colander, place in food processor with steel blade, and process until smooth.
2. Add avocado and process until desired consistency.
3. Add salsa and cilantro and pulse two or three times.

Nutritional Analysis
(per serving, excluding unknown items)

156	Calories
10 gm.	Fat
55.6%	Calories from Fat
11 gm.	Protein
7 gm.	Carbohydrate
2 gm.	Dietary Fiber
3 mg.	Cholesterol
230 mg.	Sodium

Exchanges

1½	Lean Meat
½	Fruit
2	Fat

Mayonnaise

Serves 24

If mixture breaks, place in a bowl and beat with a wire whisk vigorously.

½ teaspoon mustard
⅛ teaspoon sea salt
⅛ teaspoon white pepper
¼ cup egg substitute
1 tablespoon vinegar
1 cup olive oil

1. Mix all ingredients, except oil, in a food processor or blender.
2. With machine running, slowly add olive oil in a thin stream.
3. Keep refrigerated in a covered container for up to two weeks.

Nutritional Analysis (per serving, excluding unknown items)

84	Calories
9 gm.	Fat
98.0%	Calories from Fat
Trace	Protein
Trace	Carbohydrate
Trace	Dietary Fiber
Trace	Cholesterol
16 mg.	Sodium

Exchanges

2	Fat

Olive Cheddar Spread

Serves 6

Serve with crackers or celery.

8 ounces cottage cheese, dry curd
4 ounces cheddar cheese, shredded
6 medium green olives
2 ounces pimientos
⅛ teaspoon crushed red pepper flakes

1. Put cottage cheese and cheddar cheese in food processor with steel blade and process until smooth.
2. Place olives and pimentos in processor and pulse two or three times until chopped.
3. Place in an airtight container and refrigerate.

Nutritional Analysis (per serving, excluding unknown items)

115	Calories
7 gm.	Fat
54.3%	Calories from Fat
11 gm.	Protein
2 gm.	Carbohydrate
Trace	Dietary Fiber
22 mg.	Cholesterol
162 mg.	Sodium

Exchanges

1½	Lean Meat
1	Fat

Raspberry Sauce

Serves 6

12 ounces raspberries (frozen), thawed

2 teaspoons cornstarch

1 tablespoon sugar substitute

¼ cup apple juice, frozen concentrate

1. Combine all ingredients and cook, stirring constantly over medium heat until sauce thickens (about 10 minutes).
2. Remove from heat.
3. Serve hot or cold.

Nutritional Analysis
(per serving,
excluding unknown items)

85	Calories
Trace	Fat
1.4%	Calories from Fat
Trace	Protein
21 gm.	Carbohydrate
3 gm.	Dietary Fiber
0 mg.	Cholesterol
8 mg.	Sodium

Exchanges

1½   Fruit

Satay Dipping Sauce

Serves 6

A good dipping sauce for poultry and meat, or tossed in pasta.

⅓ cup hot water
¼ cup barbecue sauce
¼ cup peanut butter
¼ cup soy sauce
¼ cup cilantro
2 tablespoons mustard

1. Mix all ingredients together. Place in a covered container and refrigerate.

Nutritional Analysis (per serving, excluding unknown items)

85	Calories
6 gm.	Fat
58.8%	Calories from Fat
4 gm.	Protein
5 gm.	Carbohydrate
1 gm.	Dietary Fiber
0 mg.	Cholesterol
886 mg.	Sodium

Exchanges

½	Lean Meat
1	Fat

Simple Tomato Sauce

Serves 8

Sauce can be made in advance and refrigerated up to two weeks; or freeze in two-cup containers to have ready when needed.

2 teaspoons olive oil
1 medium onion, chopped
2 cloves garlic, chopped
2 28-ounce cans whole tomatoes
1 6-ounce can tomato paste
2 teaspoons dried Italian seasoning
Salt and pepper to taste

1. In a medium, heavy-bottom saucepan, stir together olive oil, onion, and garlic. Cook over low heat, stirring often, until onion is very soft, about 6 to 8 minutes.
2. Using a food processor or blender, purée tomatoes. Add tomatoes and tomato paste to onions. Bring to a boil, then reduce heat to simmer for 45 minutes, stirring bottom of sauce pan often to prevent burning.
3. Season to taste with salt and pepper.

*Nutritional Analysis
(per serving,
excluding unknown items)*

72	Calories
2 gm.	Fat
16.7%	Calories from Fat
3 gm.	Protein
14 gm.	Carbohydrate
3 gm.	Dietary Fiber
0 mg.	Cholesterol
462 mg.	Sodium

Exchanges

2½	Vegetable

Walnut Oil Mayonnaise

Serves 24

If mixture breaks, place in a bowl and beat with a wire whisk vigorously.

½ teaspoon mustard
⅛ teaspoon sea salt
⅛ teaspoon white pepper
¼ cup egg substitute
1 tablespoon balsamic vinegar
1 cup walnut oil

1. Using a food processor or blender, mix all ingredients, except walnut oil.
2. With machine running, slowly add oil in a thin steam.
3. Refrigerate in a covered container up to two weeks.

---

*Nutritional Analysis
(per serving,
excluding unknown items)*

85	Calories
9 gm.	Fat
98.0%	Calories from Fat
Trace	Protein
Trace	Carbohydrate
Trace	Dietary Fiber
Trace	Cholesterol
16 mg.	Sodium

Exchanges

2	Fat

Yogurt Cheese

Serves 4

This turns yogurt thick and creamy, add fruit or jam to it for a sweet treat or add dill, garlic, and salt to it to make a great dip for vegetables.

16 ounces plain nonfat yogurt (without any added gelatin)

1. Place yogurt in a yogurt cheese strainer or a colander lined with coffee filters.
2. Put strainer in a large bowl to catch liquid and cover top.
3. Refrigerate for 18 to 24 hours.
4. Throw out liquid and store yogurt cheese in a covered container until ready to use.

Nutritional Analysis
(per serving,
excluding unknown items)

63	Calories
Trace	Fat
2.9%	Calories from Fat
6 gm.	Protein
9 gm.	Carbohydrate
0 gm.	Dietary Fiber
2 mg.	Cholesterol
87 mg.	Sodium

Exchanges

½   Non-Fat Milk

Yogurt Cream

Serves 4

Use this on fruit or as topping on tarts and pumpkin cups.

16 ounces plain nonfat yogurt (without any added gelatin)

2 teaspoons vanilla extract

2 teaspoons sugar substitute

1. Place yogurt in a yogurt cheese strainer or a colander lined with coffee filters.
2. Put strainer in a large bowl to catch liquid and cover top.
3. Refrigerate for 18 to 24 hours.
4. Throw out liquid.
5. Add vanilla extract and sugar substitute.
6. Mix together and store in a covered container until ready to use.

Nutritional Analysis (per serving, excluding unknown items)

81	Calories
Trace	Fat
2.4%	Calories from Fat
6 gm.	Protein
12 gm.	Carbohydrate
0 gm.	Dietary Fiber
2 mg.	Cholesterol
99 mg.	Sodium

Exchanges

½	Non-Fat Milk

Breads

Applesauce Breakfast Cake
French Bread
Oat Bran Muffins
Rolled Prosciutto Sandwich
Whole Wheat Beer Bread

Applesauce Breakfast Cake

Serves 8

1 cup sugar
1 egg
¾ cup flour
¾ cup oat flour
1½ teaspoons baking soda
1 teaspoon cinnamon
¾ teaspoon nutmeg
½ teaspoon salt
½ teaspoon ground cloves
3¼ cups applesauce
½ cup raisins
½ cup chopped walnuts

1. Preheat over to 325°.
2. Mix sugar and egg.
3. Stir together flour, oat flour, baking soda, cinnamon, nutmeg, salt, and cloves.
4. Gradually add to egg mixture.
5. Mix applesauce into batter.
6. Stir in raisins and nuts.
7. Pour into greased and floured loaf pan.
8. Baked for 1 hour.
9. Cool in pan for 10 minutes. Remove from pan, cool on wire rack.

Nutritional Analysis
(per serving, excluding unknown items)

394	Calories
8 gm.	Fat
16.5%	Calories from Fat
8 gm.	Protein
78 gm.	Carbohydrate
4 gm.	Dietary Fiber
27 mg.	Cholesterol
387 mg.	Sodium

Exchanges

½	Grain (Starch)
½	Lean Meat
2	Fruit
1	Fat
1½	Other Carbohydrates

French Bread

Serves 12

1 cup warm water

1½ tablespoons yeast cake

2 cups bread flour

1 teaspoon sea salt

1 teaspoon sugar

1 teaspoon butter

1. Put all ingredients in order in bread machine and bake as directed.
2. Variations:

   Replace 1 tablespoon flour with 1 tablespoon toasted wheat germ (gives it a European style).

   Replace 3 tablespoons flour with 3 tablespoons potato buds (adds moisture).

Nutritional Analysis (per serving, excluding unknown items)

89	Calories
1 gm.	Fat
7.6%	Calories from Fat
3 gm.	Protein
17 gm.	Carbohydrate
Trace	Dietary Fiber
1 mg.	Cholesterol
162 mg.	Sodium

Exchanges

1	Grain (Starch)

Oat Bran Muffins

Yields 18 muffins

¼ cup sugar substitute, firmly packed
2½ cups oat bran, uncooked
2 tablespoons molasses
¼ cup chopped nuts
¼ cup raisins
1 tablespoon baking powder
¼ teaspoon salt (optional)
2 large eggs, lightly beaten
¾ cup skim milk
¼ cup honey
1 teaspoon almond extract
1 teaspoon vanilla extract
vegetable oil spray

1. Preheat oven to 350°F.
2. In a bowl, combine oat bran, sugar substitute, molasses, nuts, raisins, baking powder, and salt. Mix well.
3. In a small bowl, combine egg whites, milk, honey, oil, almond, and vanilla extracts.
4. Add to dry ingredients and mix to blend.
5. Spray muffin tins lightly with vegetable oil spray or use paper muffin cups.
6. Spoon mixture evenly into muffin cups.
7. Bake 20 to 25 minutes, or until light brown. Serve warm or at room temperature.

Nutritional Analysis (per serving, excluding unknown items)

134	Calories
4 gm.	Fat
21.2%	Calories from Fat
6 gm.	Protein
27 gm.	Carbohydrate
3 gm.	Dietary Fiber
36 mg.	Cholesterol
197 mg.	Sodium

Exchanges

1	Grain (Starch)
½	Fat
½	Other Carbohydrates

Rolled Prosciutto Sandwich

___
Serves 6
___

12 ounces pizza dough (Pillsbury®, for example)

1 teaspoon olive oil

4 ounces roasted red peppers, sliced ¼-inch thick

4 ounces prosciutto, thinly sliced

4 ounces Parmesan cheese, grated

1. Preheat oven to 350°F.
2. Roll out pizza dough in a rectangle and coat with olive oil.
3. Place roasted red peppers in a strip along the edge.
4. Lay prosciutto on top of dough and sprinkle with Parmesan cheese.
5. Roll dough starting with the red pepper edge.
6. Place seam side down on a baking sheet and spray dough with olive oil.
7. Bake for 25 minutes.
8. Allow bread to cool at least one hour before slicing.

Nutritional Analysis (per serving, excluding unknown items)

259	Calories
9 gm.	Fat
33.3%	Calories from Fat
17 gm.	Protein
26 gm.	Carbohydrate
1 gm.	Dietary Fiber
28 mg.	Cholesterol
863 mg.	Sodium

Exchanges

1½	Grain (Starch)
2	Lean Meat
1	Fat

Whole Wheat Beer Bread

Serves 12

1 small Spanish onion, cut in quarters
2 medium eggs
12 ounces beer, light
1½ cups bread flour
1½ cups whole wheat flour
2 teaspoons baking powder
2 teaspoons Italian seasoning, crushed
½ teaspoon red chili flakes, crushed
2 cups mozzarella cheese, grated

1. Preheat oven to 350°F.
2. Using a food processor with steel blade and ingredients in the order they are listed, pulse until mixture is well mixed.
3. Spray a Bundt® or loaf pan with nonstick spray. Fill with batter.
4. Bake for 40 to 50 minutes. Cool 10 minutes before removing from pan.

---

*Nutritional Analysis
(per serving,
excluding unknown items)*

199	Calories
6 gm.	Fat
28.0%	Calories from Fat
9 gm.	Protein
26 gm.	Carbohydrate
2 gm.	Dietary Fiber
52 mg.	Cholesterol
174 mg.	Sodium

Exchanges

1½	Grain (Starch)
½	Lean Meat
½	Fat

Salads

Almond Oil Vinaigrette Salad

Beet, Orange, and Walnut Salad

Broccoli Slaw

Caesar Salad with Roasted Walnut Oil

Chopped Feta, Tomato, and Lettuce Salad

Cucumber Salad

Fresh Spinach Salad with Pecans and Gorgonzola

Spinach Arugula Salad with Walnuts and Mandarin Oranges

Spinach, Raspberry, and Walnut Salad

Spinach Salad with Almonds

Tabbouleh

Tomato and Basil with Fresh Mozzarella

Walnut and Roquefort Cheese Salad

Almond Oil Vinaigrette Salad

Serves 4

¼ cup sherry vinegar
1 teaspoon Dijon mustard
½ teaspoon kosher salt
½ teaspoon black pepper
⅓ cup almond oil
6 cups mixed greens

1. Combine the vinegar, mustard, salt, and pepper. Whisk in the oil.
2. Toss in mixed greens.

Nutritional Analysis (per serving, excluding unknown items)

186	Calories
18 gm.	Fat
84.5%	Calories from Fat
2 gm.	Protein
5 gm.	Carbohydrate
3 gm.	Dietary Fiber
0 mg.	Cholesterol
272 mg.	Sodium

Exchanges

1	Vegetable
3½	Fat

Beet, Orange, and Walnut Salad

Serves 4

1 can beets, quartered
1 can mandarin oranges in light syrup
⅓ cup walnuts
1 tablespoon walnut oil
½ teaspoon sea salt
½ teaspoon black pepper
2 heads romaine lettuce, sliced 1-inch thick
¼ cup red wine vinegar
½ teaspoon sugar

1. Drain beets and discard liquid. Drain mandarin oranges and set aside, reserving liquid.
2. In a bowl, combine 1 teaspoon walnut oil, sugar, salt, and pepper. Add the walnuts and toss to coat.
3. Place the walnuts in a single layer on a baking dish and bake 5 to 7 minutes.
4. Allow nuts to cool and chop coarsely.
5. To make vinaigrette dressing: In a bowl, combine ½ cup mandarin orange liquid, the remaining 2 teaspoons walnut oil, red wine vinegar, and additional salt and pepper to taste.
6. Toss the lettuce, beets, and oranges with vinaigrette. Place on a serving platter or individual salad plates and sprinkle salad with nuts.
7. Serve immediately.

Nutritional Analysis
(per serving, excluding unknown items)

194	Calories
10 gm.	Fat
41.5%	Calories from Fat
9 gm.	Protein
23 gm.	Carbohydrate
8 gm.	Dietary Fiber
0 mg.	Cholesterol
283 mg.	Sodium

Exchanges

½	Lean Meat
2	Vegetable
½	Fruit
1½	Fat

Broccoli Slaw

Serves 4

⅓ cup orange juice
⅓ cup rice wine vinegar
1 tablespoon hoisin sauce
1 tablespoon soy sauce, low sodium
1 tablespoon sesame seeds, toasted
10 ounces broccoli, grated

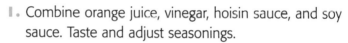

1. Combine orange juice, vinegar, hoisin sauce, and soy sauce. Taste and adjust seasonings.
2. Add grated broccoli to bowl and toss to combine.
3. Sprinkle toasted seeds on top. Serve immediately.

*Nutritional Analysis
(per serving,
excluding unknown items)*

48	Calories
1 gm.	Fat
23.7%	Calories from Fat
2 gm.	Protein
8 gm.	Carbohydrate
2 gm.	Dietary Fiber
Trace	Cholesterol
227 mg.	Sodium

Exchanges

½	Vegetable

Caesar Salad with Roasted Walnut Oil

Serves 6

For a main dish, top with grilled chicken breast or shrimp.

- 1 teaspoon anchovy paste
- 1 teaspoon garlic paste
- 1 medium lemon
- 1 tablespoon worcestershire sauce
- ¼ cup egg substitute
- 8 ounces walnut oil (such as La Tourangelle Roasted Walnut Oil)
- 3 ounces Parmesan cheese
- ¼ cup parsley
- ¼ teaspoon black pepper
- ¼ teaspoon sea salt
- 1 head romaine lettuce leaves, sliced ½-inch thick

1. To make dressing: Add anchovy paste, garlic paste, juice of the lemon, worcestershire sauce, and egg substitute into food processor or blender and blend for 15 seconds.
2. While machine is running, slowly add in small amounts of the roasted walnut oil until emulsified. Add parsley and blend one pulse.
3. Salt and pepper to taste. Toss romaine lettuce with dressing and Parmesan cheese. Serve.

Nutritional Analysis (per serving, excluding unknown items)

88	Calories
6 gm.	Fat
55.2%	Calories from Fat
8 gm.	Protein
3 gm.	Carbohydrate
Trace	Dietary Fiber
11 mg.	Cholesterol
388 mg.	Sodium

Exchanges

1	Lean Meat
½	Fat

Chopped Feta, Tomato, and Lettuce Salad

Serves 4

These flavors all blend together to create a great salad.

1 head iceberg lettuce, chopped

3 medium Roma tomatoes, seeded and chopped

3 ounces feta cheese, crumbled

½ cup rice wine vinegar

1. Mix all ingredients together and serve.

Nutritional Analysis
(per serving,
excluding unknown items)

96	Calories
5 gm.	Fat
43.4%	Calories from Fat
5 gm.	Protein
10 gm.	Carbohydrate
3 gm.	Dietary Fiber
19 mg.	Cholesterol
258 mg.	Sodium

Exchanges

½	Lean Meat
1½	Vegetable
½	Fat

Cucumber Salad

Serves 6

2 large English cucumbers, thinly sliced

3 medium scallions, thinly sliced

⅓ cup rice wine vinegar

2 teaspoons dark sesame oil

2 tablespoons lime juice

2 tablespoons ginger root, grated

1 tablespoon red chili flakes

2 tablespoons sugar substitute (such as Splenda®)

¼ cup cilantro, chopped

1. In a large bowl, combine all ingredients. Serve immediately.

Nutritional Analysis
(per serving,
excluding unknown items)

45	Calories
2 gm.	Fat
28.5%	Calories from Fat
2 gm.	Protein
7 gm.	Carbohydrate
2 gm.	Dietary Fiber
0 mg.	Cholesterol
12 mg.	Sodium

Exchanges

½   Fat

Fresh Spinach Salad with Pecans and Gorgonzola

Serves 4

10 ounces fresh spinach
½ cup pecan halves
¼ cup hazelnut oil
1 teaspoon sugar
¼ tablespoon sea salt
¼ teaspoon black pepper
1 medium shallot, chopped
½ cup rice wine vinegar
2 ounces gorgonzola cheese

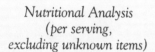

1. Preheat oven to 400°F.
2. Place spinach in serving bowl.
3. Toss pecans with 1 teaspoon hazelnut oil, sugar, and salt. Lay pecans halves on baking sheet. Roast for about 5 to 6 minutes, watching carefully. Set aside.
4. Make dressing by combining remaining hazelnut oil, vinegar, shallots, and salt and pepper in a bowl and whisking together; or place in sealable container and shake.
5. Toss spinach with dressing, gorgonzola cheese, and pecans. Serve.

---

*Nutritional Analysis
(per serving,
excluding unknown items)*

287	Calories
28 gm.	Fat
80.6%	Calories from Fat
6 gm.	Protein
9 gm.	Carbohydrate
3 gm.	Dietary Fiber
13 mg.	Cholesterol
607 mg.	Sodium

Exchanges

½	Lean Meat
½	Vegetable
5	Fat

Spinach Arugula Salad with Walnuts and Mandarin Oranges

Serves 6

For a main dish, top with grilled chicken breast or shrimp.

8 ounces mandarin oranges in light syrup

2 tablespoons rice wine vinegar

¼ cup egg substitute

8 ounces walnut oil

¼ teaspoon black pepper

¼ teaspoon sea salt

½ cup walnuts

10 ounces spinach leaves, washed and dried

10 ounces arugula leaves, washed and dried

1. Drain mandarin oranges and reserve syrup.
2. Add rice wine vinegar, salt, pepper, egg substitute, and ¼ cup mandarin syrup into food processor or blender and blend for 15 seconds.
3. While machine is running, slowly add in small amounts of walnut oil until emulsified.
4. Toss dressing, walnuts, and mandarin oranges into spinach and arugula.
5. Serve immediately.

Nutritional Analysis (per serving, excluding unknown items)

457	Calories
45 gm.	Fat
85.8%	Calories from Fat
6 gm.	Protein
11 gm.	Carbohydrate
2 gm.	Dietary Fiber
Trace	Cholesterol
140 mg.	Sodium

Exchanges

½	Lean Meat
½	Vegetable
½	Fruit
8½	Fat

Spinach, Raspberry, and Walnut Salad

Serves 6

2 pounds fresh baby spinach, washed and dried

1 cup raspberries

¼ cup walnuts, chopped

¼ cup wine vinegar

½ cup raspberry juice

¼ cup walnut oil

1. Place spinach, raspberries, and walnuts in a large salad bowl.
2. Mix together vinegar, raspberry juice, and walnut oil. Pour over spinach and toss to coat all pieces.

Nutritional Analysis
(per serving,
excluding unknown items)

157	Calories
13 gm.	Fat
65.9%	Calories from Fat
6 gm.	Protein
9 gm.	Carbohydrate
6 gm.	Dietary Fiber
0 mg.	Cholesterol
120 mg.	Sodium

Exchanges

1	Vegetable
2½	Fat

Spinach Salad with Almonds

Serves 6

10 ounces spinach leaves, washed and dried
1 cup sliced almonds
¼ cup almond oil
2 tablespoons sherry vinegar
1 tablespoon dry sherry
½ teaspoon sea salt
½ teaspoon black pepper

1. Place spinach leaves in a salad bowl.
2. Sauté almonds in almond oil until golden.
3. Toss spinach with vinegar and sherry.
4. Add almonds with oil.
5. Add salt and pepper and toss.

Nutritional Analysis (per serving, excluding unknown items)

233	Calories
22 gm.	Fat
80.4%	Calories from Fat
6 gm.	Protein
6 gm.	Carbohydrate
3 gm.	Dietary Fiber
0 mg.	Cholesterol
186 mg.	Sodium

Exchanges

½	Grain (Starch)
½	Lean Meat
4	Fat

Tabbouleh

Serves 4

½ cup bulgur
1¼ cups water
4 cups Italian parsley, chopped
1½ cups diced tomatoes
¼ cup scallions, chopped
1 medium cucumber, peeled, seeded, and sliced
⅓ cup mint leaves, chopped
¼ cup lemon juice
¼ teaspoon garlic paste
1 teaspoon kosher salt
1 tablespoon olive oil
4 cups mixed greens
¼ cup balsamic vinegar

1. Using a Dutch oven, bring water to a boil.
2. Pour in bulgur, stir, cover, and turn off heat. Let stand 20 to 25 minutes, or until most of liquid is absorbed and bulgur is fluffy and tender. Pour off any remaining liquid.
3. Prepare dressing in a small, non reactive bowl. Whisk together lemon juice, garlic paste, kosher salt, and olive oil. Taste and adjust seasonings. Set aside.
4. In a large salad bowl, toss together parsley, tomatoes, scallions, and mint. Add bulgur. Pour dressing over salad and toss to combine. Serve bulgur on top of mixed greens that have been tossed with the balsamic vinegar.

*Nutritional Analysis
(per serving,
excluding unknown items)*

131	Calories
1 gm.	Fat
6.8%	Calories from Fat
7 gm.	Protein
28 gm.	Carbohydrate
9 gm.	Dietary Fiber
0 mg.	Cholesterol
534 mg.	Sodium

Exchanges

1	Grain (Starch)
2½	Vegetable

Tomato and Basil with Fresh Mozzarella

Serves 4

2 large tomatoes, sliced ½-inch thick
8 ounces mozzarella cheese, sliced
fresh basil leaves
¼ cup balsamic vinegar

1. Alternate tomato slices, mozzarella cheese slices, and basil leaves and drizzle balsamic vinegar over the top.

Nutritional Analysis
(per serving,
excluding unknown items)

196	Calories
14 gm.	Fat
63.9%	Calories from Fat
13 gm.	Protein
5 gm.	Carbohydrate
1 gm.	Dietary Fiber
51 mg.	Cholesterol
241 mg.	Sodium

Exchanges

2	Lean Meat
½	Vegetable
2	Fat

Walnut and Roquefort Cheese Salad

Serves 6

2 heads Belgian endive, sliced 1-inch thick
1 head romaine lettuce, sliced 1-inch thick
1½ tablespoons rice wine vinegar
¼ cup egg substitute
½ teaspoon sea salt
½ teaspoon black pepper
6 tablespoons olive oil
2 whole pears
3 ounces Roquefort cheese, crumbled
½ cup walnuts, chopped

1. Place Belgian endive and romaine lettuce in a large salad bowl.
2. In a medium bowl, whisk together rice wine vinegar, egg substitute, sea salt, and black pepper.
3. Slowly whisk in olive oil until emulsified.
4. Toss pears with some vinaigrette and place on lettuce mixture.
5. Drizzle remaining vinaigrette over lettuce and pears.
6. Add the crumbled Roquefort cheese and walnuts.
7. Toss and serve at room temperature

Nutritional Analysis (per serving, excluding unknown items)

306	Calories
25 gm.	Fat
70.7%	Calories from Fat
9 gm.	Protein
14 gm.	Carbohydrate
5 gm.	Dietary Fiber
13 mg.	Cholesterol
444 mg.	Sodium

Exchanges

1	Lean Meat
½	Vegetable
½	Fruit
4½	Fat

Side Dishes

Apricot and Date Risotto
Asparagus with Walnuts
Baked Cauliflower
Broccoli Florets
Brown Rice with Cashews
Brown Rice with Spicy Pecans
Couscous with Black Beans
Couscous with Mint
Couscous, Peas, and Carrots
Couscous, Plain and Simple
Couscous and Zucchini
Fresh Mozzarella and Tomatoes
Red Pepper Couscous
Salmon Cucumber Slices
Smoked Salmon Dip
Snow Peas
Spinach Goat Cheese Toast
Spinach Potato Pie
Tapenade
Whole Wheat Cheese Quesadilla
Zucchini and Carrots Julienne

Apricot and Date Risotto

Serves 6

3 tablespoons olive oil
3 tablespoons hot sauce
1 medium onion, chopped
1½ cups rice, short-grain, brown or white
½ cup vermouth
5 cups low sodium chicken broth
½ cup parsley, chopped
1 cup dried apricot halves, chopped
½ cup walnuts, chopped

1. Heat oil with hot sauce and cook the onion in it until soft.
2. Add rice and stir over medium-high heat until all rice is coated with oil.
3. Add vermouth and stir until absorbed.
4. Add chicken broth, one cup at a time, until the rice absorbs liquid. (Have extra water ready in case you need it.)
5. Cook until rice is tender, stirring constantly. (Brown rice takes about 15 minutes longer to cook than white rice.)
6. Add parsley, date nuggets, chopped apricots, and walnuts. Stir in well.

Nutritional Analysis
(per serving, excluding unknown items)

477	Calories
13 gm.	Fat
25.4%	Calories from Fat
17 gm.	Protein
70 gm.	Carbohydrate
6 gm.	Dietary Fiber
0 mg.	Cholesterol
632 mg.	Sodium

Exchanges

2½	Grain (Starch)
1½	Lean Meat
½	Vegetable
1½	Fruit
2½	Fat

Asparagus with Walnuts

Serves 4

1 pound asparagus spears
4 teaspoons red wine vinegar
1 tablespoon walnut oil
¼ teaspoon sea salt
¼ teaspoon black pepper
¼ cup walnuts, chopped

1. Preheat oven to 400°F.
2. Bend each asparagus until it breaks, throwing out the end section.
3. Cut the asparagus in ½-inch pieces on an angle.
4. On a sheet pan, toss asparagus with red wine vinegar, walnut oil, sea salt, and black pepper.
5. Roast in a preheated oven for 10 minutes.
6. Serve topped with chopped walnuts.

Nutritional Analysis (per serving, excluding unknown items)

92	Calories
8 gm.	Fat
71.0%	Calories from Fat
3 gm.	Protein
4 gm.	Carbohydrate
2 gm.	Dietary Fiber
0 mg.	Cholesterol
119 mg.	Sodium

Exchanges

½	Vegetable
1½	Fat

Baked Cauliflower

Serves 4

1 pound frozen cauliflower florets, thawed
½ teaspoon sea salt
½ teaspoon black pepper
½ teaspoon garlic powder
1 teaspoon dried oregano
½ cup dry bread crumbs
2 tablespoons fresh parsley
¼ cup Parmesan cheese

1. Preheat the oven to 400°F.
2. Place cauliflower in a baking dish that has been sprayed with a cooking spray.
3. Mix all other ingredients together, and spread over the florets, bake uncovered in oven until the topping is browned, about 15 minutes. Serve hot or at room temperature.

Nutritional Analysis
(per serving,
excluding unknown items)

107	Calories
3 gm.	Fat
20.8%	Calories from Fat
6 gm.	Protein
16 gm.	Carbohydrate
3 gm.	Dietary Fiber
4 mg.	Cholesterol
473 mg.	Sodium

Exchanges

½	Grain (Starch)
½	Lean Meat
1	Vegetable
½	Fat

Broccoli Florets

Serves 4

1 pound frozen broccoli florets, thawed
½ teaspoon sea salt
½ teaspoon black pepper
½ teaspoon garlic powder
1 teaspoon Italian seasoning
½ cup dry bread crumbs
2 tablespoons fresh parsley
¼ cup Parmesan cheese

1. Preheat the oven to 400°F.
2. Place broccoli florets in a baking dish that has been sprayed with a cooking spray.
3. Mix all other ingredients together, and spread over the florets. Bake uncovered in oven until the topping is browned, about 15 minutes. Serve hot or at room temperature.

Nutritional Analysis
(per serving, excluding unknown items)

79	Calories
2 gm.	Fat
26.0%	Calories from Fat
4 gm.	Protein
11 gm.	Carbohydrate
1 gm.	Dietary Fiber
4 mg.	Cholesterol
446 mg.	Sodium

Exchanges

½	Grain (Starch)
½	Lean Meat

Brown Rice with Cashews

Serves 4

2¼ cups chicken broth
1 cup brown rice
1 teaspoon ginger root, grated
2 whole scallions, chopped
½ cup cashews

1. Bring chicken broth to boil, add brown rice, ginger root, and scallions.
2. Reduce heat to simmer, cover, and cook 45 to 50 minutes.
3. Add cashews into rice. Season to taste. Serve.

---

*Nutritional Analysis
(per serving,
excluding unknown items)*

290	Calories
10 gm.	Fat
30.4%	Calories from Fat
9 gm.	Protein
42 gm.	Carbohydrate
2 gm.	Dietary Fiber
0 mg.	Cholesterol
435 mg.	Sodium

Exchanges

2½	Grain (Starch)
½	Lean Meat
1½	Fat

Brown Rice with Spicy Pecans

Serves 8

4¼ cups chicken broth
½ teaspoon sea salt
¼ teaspoon black pepper
2 cups brown rice
3 whole scallions, chopped
1 cup pecan halves
1 tablespoon butter, melted
⅛ teaspoon cayenne pepper
1 tablespoon worcestershire sauce
¼ teaspoon hot sauce
⅛ teaspoon cinnamon
½ teaspoon sea salt

1. Bring the chicken stock to a boil. Add the salt, pepper, rice, and scallions.
2. Return to a boil, cover, reduce to a simmer, and cook until the rice is tender, approximately 45 to 50 minutes.
3. Combine the butter, cayenne pepper, worcestershire sauce, hot sauce, cinnamon, and salt. Add the pecan halves and blend well.
4. Spread the nuts onto a sheet pan and bake at 300°F for 10 minutes.
5. Toss with a spatula and bake for an additional 5 to 10 minutes. Remove from the oven and cool.
6. Chop the nuts and mix in brown rice. Serve.

Nutritional Analysis
(per serving, excluding unknown items)

299	Calories
13 gm.	Fat
37.4%	Calories from Fat
7 gm.	Protein
40 gm.	Carbohydrate
2 gm.	Dietary Fiber
4 mg.	Cholesterol
680 mg.	Sodium

Exchanges

2½	Grain (Starch)
½	Lean Meat
2	Fat

Couscous with Black Beans

Serves 8

1 tablespoon olive oil
1 scallion, chopped
1 whole red bell pepper, diced into ¼-inch pieces
10½ ounces canned black beans, drained and rinsed
2 cups vegetable broth
⅛ cup fresh lime juice
2 cups couscous
¼ cup parsley, chopped

1. Using a Dutch oven, sauté onion and bell pepper just until soft.
2. Stir beans, vegetable broth, and lime juice into sautéed vegetables. Bring to a rolling boil.
3. Add couscous to liquid and stir. Remove from heat.
4. Cover and let stand five minutes. Fluff with fork before serving.
5. Stir parsley into couscous.

Nutritional Analysis
(per serving,
excluding unknown items)

393	Calories
14 gm.	Fat
30.8%	Calories from Fat
20 gm.	Protein
52 gm.	Carbohydrate
5 gm.	Dietary Fiber
0 mg.	Cholesterol
10886 mg.	Sodium

Exchanges

2½	Grain (Starch)
½	Fat

Couscous with Mint

Serves 8

1 tablespoon olive oil
1 small onion, chopped
½ cup carrots, chopped
1 clove garlic, chopped
2 cups couscous
2 cups chicken broth
Black pepper and sea salt to taste
2 tablespoons mint leaves, chopped

1. In a large saucepan or a wok, heat the olive oil.
2. Sauté onion, garlic, and carrots for 2 or 3 minutes.
3. Add the couscous and coat well.
4. Pour in the chicken broth and cook for 2 minutes, stirring occasionally.
5. Stir in the mint and mix thoroughly.
6. Season with salt and pepper to taste and keep warm.
7. Fluff with fork before serving.

Nutritional Analysis
(per serving,
excluding unknown items)

197	Calories
2 gm.	Fat
10.9%	Calories from Fat
7 gm.	Protein
36 gm.	Carbohydrate
3 gm.	Dietary Fiber
0 mg.	Cholesterol
199 mg.	Sodium

Exchanges

2	Grain (Starch)
½	Vegetable
½	Fat

Couscous, Peas, and Carrots

Serves 8

2 cups low sodium chicken broth
2 cups couscous
¼ teaspoon black pepper
10 ounces frozen peas and carrots, thawed

1. Bring chicken broth and pepper to boil.
2. Add couscous, peas, and carrots; cover and remove from heat.
3. Let stand for five minutes. Fluff with fork before serving.

Nutritional Analysis
(per serving, excluding unknown items)

194	Calories
Trace	Fat
2.0%	Calories from Fat
9 gm.	Protein
38 gm.	Carbohydrate
3 gm.	Dietary Fiber
0 mg.	Cholesterol
162 mg.	Sodium

Exchanges

2½	Grain (Starch)
½	Lean Meat

Couscous, Plain and Simple

Serves 8

2 cups low sodium chicken broth
2 cups couscous
¼ teaspoon black pepper

1. Bring chicken broth and pepper to boil.
2. Stir couscous, cover, and remove from heat.
3. Let stand for five minutes. Fluff with fork before serving.

Nutritional Analysis
(per serving, excluding unknown items)

175	Calories
Trace	Fat
1.5%	Calories from Fat
8 gm.	Protein
34 gm.	Carbohydrate
2 gm.	Dietary Fiber
0 mg.	Cholesterol
134 mg.	Sodium

Exchanges

2	Grain (Starch)
½	Lean Meat

Couscous and Zucchini

Serves 6

1½ cups chicken broth
1 cup couscous
½ cup red bell pepper, chopped fine
2 teaspoons olive oil
2 medium zucchini or yellow squash, sliced ¼-inch thick
1 teaspoon red chili flakes
½ teaspoon garlic powder
2 tablespoons fresh basil, chopped fine
1 tablespoon balsamic vinegar
1 teaspoon sea salt

1. Bring chicken broth to a boil. Add couscous and red pepper and cover.
2. Remove from heat and let set for 5 minutes.
3. Heat a medium sauté pan. Add olive oil, zucchini or squash, and remaining ingredients and continue to cook until tender.
4. Mound couscous in center of a large platter. Arrange zucchini or squash slices around the couscous. Serve warm or cold.

Nutritional Analysis (per serving, excluding unknown items)

139	Calories
2 gm.	Fat
10.9%	Calories from Fat
7 gm.	Protein
26 gm.	Carbohydrate
3 gm.	Dietary Fiber
0 mg.	Cholesterol
443 mg.	Sodium

Exchanges

1½	Grain (Starch)
½	Lean Meat
½	Vegetable
½	Fat

Fresh Mozzarella and Tomatoes

Serves 4

4 medium tomatoes
8 ounces mozzarella cheese
12 basil leaves
2 tablespoons balsamic vinegar
½ teaspoon sea salt
½ teaspoon black pepper

1. Slice the tomatoes and mozzarella.
2. Arrange casually with the basil leaves on a large platter, or arrange on 4 individual salad plates.
3. Drizzle with balsamic vinegar.
4. Sprinkle with salt and pepper and serve at room temperature.

Nutritional Analysis
(per serving, excluding unknown items)

209	Calories
14 gm.	Fat
60.4%	Calories from Fat
13 gm.	Protein
8 gm.	Carbohydrate
1 gm.	Dietary Fiber
51 mg.	Cholesterol
482 mg.	Sodium

Exchanges

2	Lean Meat
1	Vegetable
2	Fat

Red Pepper Couscous

Serves 6

1 teaspoon olive oil
1 green onion, sliced
1 red bell pepper, julienne
1 clove garlic, chopped
1 teaspoon paprika
2 cups water
2 tablespoons lemon juice
1 tablespoon tomato paste
1 ounce pimiento, chopped
1/4 teaspoon red pepper flakes
1/4 teaspoon sea salt
2 cups couscous, instant

1. Using a Dutch oven, heat oil and sauté green onion, red bell pepper, and garlic until tender.
2. Add paprika, water, lemon juice, tomato paste, pimiento, red pepper flakes, and sea salt to sautéed vegetable mixture. Bring to a rolling boil.
3. Add couscous to mixture and stir. Cover and turn off heat. Let stand five minutes. Stir to fluff.

Nutritional Analysis (per serving, excluding unknown items)

236	Calories
1 gm.	Fat
4.8%	Calories from Fat
8 gm.	Protein
48 gm.	Carbohydrate
4 gm.	Dietary Fiber
0 mg.	Cholesterol
109 mg.	Sodium

Exchanges

3	Grain (Starch)
1/2	Vegetable

Salmon Cucumber Slices

Serves 4

8 slices toast
4 ounces light cream cheese
8 ounces smoked salmon
½ English cucumber, cut in ¼-inch thick slices
¼ cup lemon juice

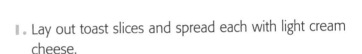

1. Lay out toast slices and spread each with light cream cheese.
2. Divide smoked salmon equally among the slices of toast.
3. Place cucumber slices on top of smoked salmon.
4. Sprinkle with lemon juice.

Nutritional Analysis
(per serving,
excluding unknown items)

141	Calories
7 gm.	Fat
47.9%	Calories from Fat
14 gm.	Protein
4 gm.	Carbohydrate
1 gm.	Dietary Fiber
29 mg.	Cholesterol
604 mg.	Sodium

Exchanges

2	Lean Meat
½	Fat

Smoked Salmon Dip

Serves 4

8 ounces light cream cheese
½ cup light sour cream
1 teaspoon dill
1 teaspoon horseradish
¼ teaspoon sea salt
¼ teaspoon white pepper
¼ cup smoked salmon

1. Cream the cheese. Add sour cream, dill, horseradish, salt, and pepper. Mix well.
2. Add smoked salmon and mix well.
3. Chill and serve with toast points (toast cut in triangles), crackers, or crudités (raw vegetables).

Nutritional Analysis (per serving, excluding unknown items)

153	Calories
11 gm.	Fat
64.2%	Calories from Fat
8 gm.	Protein
5 gm.	Carbohydrate
Trace	Dietary Fiber
36 mg.	Cholesterol
513 mg.	Sodium

Exchanges

1	Lean Meat
1½	Fat
½	Other Carbohydrates

Snow Peas

Serves 4

1 pound snow pea pods, fresh, whole
1 teaspoon grapeseed oil
1 teaspoon dark sesame oil
1 clove garlic, chopped
1 teaspoon ginger root, chopped
8 ounces water chestnuts canned and sliced

1. Clean snow peas, trimming if necessary.
2. In large skillet, heat grapeseed and sesame oils over medium heat.
3. Stir in snow peas, garlic, water chestnuts, and ginger. Sauté for 1 to 2 minutes until snow peas are crisp tender. Serve immediately.

Nutritional Analysis (per serving, excluding unknown items)

68 Calories
2 gm. Fat
32.0% Calories from Fat
3 gm. Protein
9 gm. Carbohydrate
3 gm. Dietary Fiber
0 mg. Cholesterol
5 mg. Sodium

Exchanges

1½ Vegetable
½ Fat

Spinach Goat Cheese Toast

___
Serves 6
___

6 slices bread, toasted

3 ounces goat cheese

4 ounces spinach leaves

1 medium tomato, diced

1. Lay out toast slices and spread with goat cheese.
2. Place spinach and diced tomato on top of goat cheese, and serve.

Nutritional Analysis
(per serving,
excluding unknown items)

138	Calories
6 gm.	Fat
39.3%	Calories from Fat
7 gm.	Protein
14 gm.	Carbohydrate
1 gm.	Dietary Fiber
15 mg.	Cholesterol
196 mg.	Sodium

Exchanges

1	Grain (Starch)
1/2	Lean Meat
1/2	Vegetable
1	Fat

Spinach Potato Pie

Serves 6

This is a great recipe to experiment with ... you can substitute any vegetable for the spinach. The baking time may vary so check for tenderness of the vegetables.

3 medium russet potatoes, sliced 1/8-inch thick
10 ounces frozen spinach, chopped
3 medium green onions, chopped
1 cup chicken broth
6 ounces Parmesan cheese, grated
1 teaspoon sea salt
1 teaspoon fresh ground white pepper
2 tablespoons parsley, chopped fine

1. Preheat oven to 350°F.
2. Arrange two layers of potato slices around a 9-inch quiche or pie pan that has been sprayed with a nonstick spray.
3. Spread thawed spinach and onions over potatoes. Sprinkle with 1/2 teaspoon each of sea salt and white pepper.
4. Layer remaining potato slices on top and sprinkle with remaining salt and pepper.
5. Pour chicken broth over top, cover, and bake for one hour.
6. Uncover, spread with Parmesan cheese and bake 10 to 15 minutes longer.
7. Let stand 10 minutes. Sprinkle with parsley. Cut into pie shaped wedges and serve.

Nutritional Analysis
(per serving, excluding unknown items)

181	Calories
9 gm.	Fat
44.0%	Calories from Fat
15 gm.	Protein
11 gm.	Carbohydrate
2 gm.	Dietary Fiber
22 mg.	Cholesterol
1007 mg.	Sodium

Exchanges

1/2	Grain (Starch)
1 1/2	Lean Meat
1/2	Vegetable
1/2	Fat

Tapenade

Serves 6

8 ounces black olives, pitted
3 tablespoons capers
1 teaspoon anchovy paste
1 teaspoon garlic paste
½ cup olive oil
1 tablespoon lemon juice
2 teaspoons Dijon mustard
½ teaspoon thyme
1 tablespoon fresh parsley
1 large baguette
olive oil spray

1. Combine all ingredients (except the baguette and olive oil spray) in food processor and process until chunky.
2. Slice baguette in ½-inch slices, spray with olive oil spray, and toast.
3. Serve tapenade on baguette toast.

Nutritional Analysis
(per serving,
excluding unknown items)

415	Calories
24 gm.	Fat
52.7%	Calories from Fat
7 gm.	Protein
42 gm.	Carbohydrate
4 gm.	Dietary Fiber
0 mg.	Cholesterol
850 mg.	Sodium

Exchanges

2½	Grain (Starch)
5	Fat

Whole Wheat Cheese Quesadilla

Serves 4

4 medium, whole wheat tortillas
3 ounces cheddar cheese
2 tablespoons salsa

1. Spray a 10-inch skillet with nonstick spray and bring to medium heat.
2. Place one tortilla in skillet.
3. Cover tortilla with 1½ ounces cheese and 1 tablespoon salsa.
4. Cover with another tortilla. Spray top of tortilla with nonstick spray.
5. When brown, flip over and cook other side, browning each side.
6. Repeat directions above for additional cheese quesadillas.
7. Cut each completed quesadilla into four slices and serve.

Nutritional Analysis (per serving, excluding unknown items)

228	Calories
10 gm.	Fat
37.3%	Calories from Fat
9 gm.	Protein
27 gm.	Carbohydrate
2 gm.	Dietary Fiber
22 mg.	Cholesterol
547 mg.	Sodium

Exchanges

½	Lean Meat
1	Fat

Zucchini and Carrots Julienne

Serves 4

- 2 medium zucchini, julienned
- 2 medium carrots, julienned
- 2 teaspoons butter
- 1 teaspoon olive oil
- ½ teaspoon sea salt
- ½ teaspoon black pepper

1. Heat butter and olive oil in sauce pan.
2. Add zucchini, carrots, sea salt, and black pepper; cook until crisp tender.

Nutritional Analysis (per serving, excluding unknown items)

57	Calories
3 gm.	Fat
47.1%	Calories from Fat
2 gm.	Protein
7 gm.	Carbohydrate
2 gm.	Dietary Fiber
5 mg.	Cholesterol
270 mg.	Sodium

Exchanges

1	Vegetable
½	Fat

Main Dishes—Salads

Asian Chicken Salad
Beef Sirloin Salad with Dried Cherries
Blackened Flank Steak Salad
Burgers with Grilled Onions and Salad
Scallop Salad
Steak, Tomato, and Pepper Salad

Asian Chicken Salad

Serves 4

1 head iceberg lettuce, cut in 1-inch pieces

2 cups poached chicken, cut into chunks

4 ounces spinach leaves, stems removed and sliced into strips

½ cup romaine lettuce, cut in 1-inch pieces

½ whole red bell pepper, ribs removed and cut into strips

2 scallions, sliced

3 tablespoons rice wine vinegar

1 tablespoon reduced-sodium soy sauce

1 tablespoon sesame oil

1 teaspoon Teriyaki sauce

1½ tablespoons prepared chili sauce

1 tablespoon fresh grated ginger

3 tablespoons slivered almonds, toasted

Nutritional Analysis (per serving, excluding unknown items)

365	Calories
11 gm.	Fat
27.1%	Calories from Fat
30 gm.	Protein
37 gm.	Carbohydrate
3 gm.	Dietary Fiber
60 mg.	Cholesterol
227 mg.	Sodium

Exchanges

2	Grain (Starch)
3	Lean Meat
½	Vegetable
1½	Fat

1. In a large salad bowl, combine lettuce, chicken, spinach, sprouts, pepper, and scallions.
2. In a small mixing bowl, mix together remaining ingredients except almonds; whisk well.
3. Toss dressing with salad mixture and refrigerate until ready to serve. Sprinkle almonds over top just before serving.

Beef Sirloin Salad with Dried Cherries

Serves 4

1 pound sirloin steak, trimmed, sliced ½-inch thick
8 cups mixed greens
4 ounces crumbled blue cheese
½ cup dried cherries
½ cup pine nuts
½ cup olive oil
¼ cup red wine vinegar
2 cloves garlic, minced
¾ teaspoon kosher salt
¾ teaspoon black pepper

1. Pat dry the sirloin slices.
2. Whisk dressing ingredients in a medium bowl until blended.
3. Remove ½ cup of the dressing and set aside.
4. Add beef to remaining dressing; toss to coat. Cover and marinate in refrigerator 30 minutes.
5. Remove beef from marinade; discard marinade.
6. Heat large nonstick skillet over medium-high heat until hot.
7. Add half of the beef; stir fry one to two minutes or until outside surface of beef is no longer pink. Remove from skillet. Repeat with remaining beef.
8. Combine lettuce and reserved dressing in large bowl; toss to coat.
9. Divide among four plates. Arrange beef evenly on lettuce; sprinkle with blue cheese, cherries, and nuts.

Nutritional Analysis
(per serving, excluding unknown items)

760	Calories
60 gm.	Fat
69.2%	Calories from Fat
35 gm.	Protein
25 gm.	Carbohydrate
6 gm.	Dietary Fiber
92 mg.	Cholesterol
838 mg.	Sodium

Exchanges

4	Lean Meat
1	Vegetable
1	Fruit
9	Fat

Blackened Flank Steak Salad

Serves 4

1 pound flank steak
1 tablespoon grapeseed oil
1 teaspoon garlic powder
1 teaspoon cayenne
½ teaspoon white pepper
½ teaspoon black pepper
½ teaspoon dried thyme
12 medium cherry tomatoes
8 cups mixed greens
½ cup red wine vinegar

1. Trim all excess fat off flank steak.
2. Rub the steak on both sides with grapeseed oil.
3. Mix all the spices together and spread on both sides of the steak; then press or pound the spices into the steak.
4. Cook steak on a hot grill for 4 to 5 minutes on each side.
5. Take steak off the grill and let rest for 10 minutes. (The steak will continue to cook while it rests.) Then slice steak on the bias.
6. Cut cherry tomatoes in half and toss with mixed greens and red wine vinegar.
7. Serve salad on individual plates topped with the steak strips.

---

*Nutritional Analysis
(per serving,
excluding unknown items)*

280	Calories
16 gm.	Fat
49.2%	Calories from Fat
26 gm.	Protein
11 gm.	Carbohydrate
5 gm.	Dietary Fiber
58 mg.	Cholesterol
112 mg.	Sodium

Exchanges

3	Lean Meat
1½	Vegetable
1½	Fat

Burgers with Grilled Onions and Salad

Serves 4

1 pound lean ground beef
1 large sweet onion, cut in ½-inch thick slices
½ teaspoon kosher salt
½ teaspoon black pepper
4 cups mixed greens
2 whole tomatoes, chopped
½ cup light Italian salad dressing

1. Lightly shape ground beef into 4¾-inch thick patties, sprinkle with salt and pepper. Spray onion slices with oil. Cook burgers and onions on the grill until meat reaches 160°F.
2. Toss mixed greens and tomatoes with dressing. Place the mixed salad on four plates and top each plate with burger and grilled onions.

Nutritional Analysis (per serving, excluding unknown items)

370	Calories
27 gm.	Fat
65.2%	Calories from Fat
22 gm.	Protein
10 gm.	Carbohydrate
3 gm.	Dietary Fiber
87 mg.	Cholesterol
570 mg.	Sodium

Exchanges

3	Lean Meat
1½	Vegetable
3½	Fat

Scallop Salad

Serves 4

20 ounces scallops

12 ounces romaine lettuce, sliced ½-inch thick

8 ounces baby lettuce leaves

4 ounces gorgonzola cheese, crumbled

4 ounces dried cranberries

2 ounces balsamic vinegar

1 teaspoon olive oil spray

salt and white pepper

1 lemon, sliced

1. Lightly spray scallops with olive oil and sprinkle with salt and white pepper. Grill lightly on medium-hot grill, turning as needed (do not burn or char) for approximately 2 to 3 minutes.

2. Meanwhile, place lettuce, gorgonzola cheese, cranberries, and balsamic vinegar in a mixing bowl. Toss gently with tongs.

3. Neatly place salad in middle of plate. Arrange scallops around top and sides of greens. Garnish with lemon slices.

Nutritional Analysis
(per serving,
excluding unknown items)

274	Calories
12 gm.	Fat
36.5%	Calories from Fat
32 gm.	Protein
14 gm.	Carbohydrate
3 gm.	Dietary Fiber
72 mg.	Cholesterol
633 mg.	Sodium

Exchanges

4½	Lean Meat
½	Vegetable
½	Fruit
1½	Fat

Steak, Tomato, and Pepper Salad

Serves 4

24 ounces sirloin steak, trimmed
1 teaspoon garlic paste
½ teaspoon kosher salt
½ teaspoon black pepper
2 heads romaine lettuce, sliced ½-inch thick
2 medium cucumbers, seeded and peeled
2 whole red bell peppers, julienned
1 package cherry tomatoes
¼ cup rice wine vinegar
1 medium lime
3 tablespoons olive oil

1. Preheat grill to medium-high heat.
2. Rub sirloin steak with garlic paste and sprinkle with kosher salt and black pepper. Grill medium rare, about five minutes on each side, turning occasionally to avoid burning. After meat is done, set aside and allow to rest. (The steak will continue to cook as it rests.)
3. Slice cucumbers in a large salad bowl and toss with romaine lettuce, red peppers, and cherry tomatoes. Mix rice wine vinegar, the juice of one lime, and 3 tablespoons olive oil. Salt and pepper taste.
4. Slice steak thinly into strips. Assemble salad on plate and cover with sliced steak. Pour dressing over salad. Serve immediately, family style.

Nutritional Analysis
(per serving, excluding unknown items)

479	Calories
34 gm.	Fat
63.3%	Calories from Fat
33 gm.	Protein
11 gm.	Carbohydrate
3 gm.	Dietary Fiber
107 mg.	Cholesterol
328 mg.	Sodium

Exchanges

4½	Lean Meat
1½	Vegetable
4	Fat

Main Dishes—Seafood

Baked Halibut Fillets

Brown-Bagging-It Halibut

Grilled Southern Comfort® Shrimp à l'Orange

Halibut Fillets with California Vegetables in Packets

Halibut with Vegetables in Packets

Salmon with Asparagus in Packets

Salmon Oriental in Packets

Sautéed Sea Bass

Scallops with Orzo in Packets

Scallops in White Wine

Sea Bass Oriental in Packets

Sea Bass and Vegetables in Packets

Baked Halibut Fillets

Serves 4

1 tablespoon olive oil
1 whole green bell pepper, thinly sliced
1 clove garlic, minced
1 cup diced tomatoes
1 pound halibut fillets
½ teaspoon sea salt
½ teaspoon white pepper
2 tablespoons lemon juice

1. Preheat oven to 350°F.
2. Heat olive oil in skillet; add green bell pepper and garlic. Cook until soft, add tomatoes, and stir.
3. Place halibut in a baking dish and sprinkle with salt and lemon juice.
4. Cover halibut with vegetable mixture and bake uncovered for 20 minutes or until halibut flakes.

Nutritional Analysis (per serving, excluding unknown items)

176	Calories
6 gm.	Fat
32.1%	Calories from Fat
24 gm.	Protein
5 gm.	Carbohydrate
1 gm.	Dietary Fiber
36 mg.	Cholesterol
301 mg.	Sodium

Exchanges

3½	Lean Meat
1	Vegetable
½	Fat

Brown-Bagging-It Halibut

Serves 4

1 pound halibut fillets
½ teaspoon sea salt
½ teaspoon white pepper
2 tablespoons soy sauce, low sodium
1 tablespoon hoisin sauce
2 tablespoons lemon juice
2 tablespoons ginger
4 brown paper lunch bags, sprayed with cooking oil

1. Preheat oven to 425°F.
2. Spray oil over the outside of each bag until all surfaces are coated.
3. Rinse fillets and pat dry. Season both sides of fillets with salt and pepper.
4. In a small bowl, mix soy sauce, hoisin sauce, lemon juice, and ginger.
5. Set bags on their broad side and place one fillet flat inside each bag. Then, using a tablespoon, reach into the bag and spoon one quarter of the soy-hoisin-lemon-ginger mixture over each of the fillets. Force excess air from the bags, roll up the open ends and tightly crimp to seal shut.
6. Bake on a cookie sheet for 10 to 12 minutes.
7. To serve, tear a small slit in the bags, peel back the paper just enough to expose the fish. Serve immediately.

Nutritional Analysis
(per serving,
excluding unknown items)

151	Calories
3 gm.	Fat
18.0%	Calories from Fat
25 gm.	Protein
5 gm.	Carbohydrate
1 gm.	Dietary Fiber
36 mg.	Cholesterol
662 mg.	Sodium

Exchanges

3½	Lean Meat

Grilled Southern Comfort® Shrimp à l'Orange

Serves 6

If fresh pineapple is not available, you can use canned or frozen.

½ cup Southern Comfort®
2 cups orange marmalade
½ cup pineapple juice
2 teaspoons hot sauce
1 teaspoon red chili flakes
1 teaspoon ginger root, grated
½ cup lemon juice
½ teaspoon sea salt
¼ teaspoon black pepper
2 pounds shrimp, large, shells removed
1 large pineapple, cubed
1 large onion, cubed

1. Mix all ingredients except for shrimp, pineapple, and onion in a sauce pan and bring to boil. Turn off heat and let sit.
2. Alternate shrimp on skewers with chunks of fresh pineapple and cubed onion. Grill until the shrimp turns pink and is firm to the touch.

Nutritional Analysis
(per serving, excluding unknown items)

549	Calories
3 gm.	Fat
5.1%	Calories from Fat
32 gm.	Protein
90 gm.	Carbohydrate
7 gm.	Dietary Fiber
230 mg.	Cholesterol
484 mg.	Sodium

Exchanges

4½	Lean Meat
½	Vegetable
1	Fruit
4½	Other Carbohydrates

Halibut Fillets with California Vegetables in Packets

Serves 4

- 24 ounces halibut fillets
- 16 ounces frozen California-blend vegetables
- 1 teaspoon Italian seasoning
- ½ teaspoon sea salt
- ½ teaspoon lemon pepper
- 4 sheets of aluminum foil (12x18 inches each)

1. Preheat oven to 450°F or preheat grill to medium-high.
2. Center one halibut fillet on each sheet of aluminum foil. Sprinkle fillets with salt and lemon pepper. Top with vegetables and sprinkle with Italian seasoning.
3. Bring up foil sides. Double fold top and ends to seal packet, leaving room for heat circulation inside. Repeat to make four packets.
4. Bake 16 to 20 minutes on a cookie sheet in oven or grill 12 to 16 minutes on a covered grill.

Nutritional Analysis
(per serving, excluding unknown items)

229	Calories
4 gm.	Fat
16.4%	Calories from Fat
38 gm.	Protein
7 gm.	Carbohydrate
4 gm.	Dietary Fiber
54 mg.	Cholesterol
403 mg.	Sodium

Exchanges

5	Lean Meat
1	Vegetable

Halibut with Vegetables in Packets

Serves 4

1 large lemon, thinly sliced
1½ pounds halibut fillets
½ medium zucchini, julienned
½ medium yellow squash, julienned
1 medium carrot, julienned
1 medium onion, sliced ¼-inch thick
½ teaspoon sea salt
½ teaspoon black pepper
1 teaspoon dried dill weed
4 sheets of aluminum foil (12x18 inches each)

1. Preheat oven to 450°F or preheat grill to medium-high.
2. Center several lemon slices on each sheet of foil. Place halibut on lemon slices. Sprinkle lightly with salt and pepper. Top with zucchini, yellow squash, carrot, and onion. Sprinkle dill weed over halibut and vegetables.
3. Bring up foil sides. Double fold top and ends to seal packet, leaving room for heat circulation inside. Repeat to make four packets.
4. Bake 16 to 20 minutes on a cookie sheet in oven, or grill 12 to 16 minutes on a covered grill.

---

*Nutritional Analysis
(per serving,
excluding unknown items)*

216	Calories
4 gm.	Fat
17.3%	Calories from Fat
37 gm.	Protein
7 gm.	Carbohydrate
2 gm.	Dietary Fiber
54 mg.	Cholesterol
336 mg.	Sodium

Exchanges

5	Lean Meat
1	Vegetable

Salmon with Asparagus in Packets

Serves 4

¼ cup honey
2 tablespoons Dijon mustard
1½ tablespoons melted butter
2 teaspoons worcestershire sauce
1 tablespoon cornstarch
⅛ teaspoon white pepper
1 pound asparagus
1 pound salmon fillets
4 sheets of aluminum foil (12x18 inches each)

1. Preheat oven to 450°F or grill to medium-high.
2. Blend honey, Dijon mustard, butter, worcestershire sauce, cornstarch, and white pepper; set aside.
3. Bend each asparagus until it breaks, throwing out the end section. Cut the end of the asparagus on an angle. Center one-fourth of asparagus on each sheet of foil. Top with salmon fillet; drizzle with reserved honey-mustard sauce.
4. Bring up foil sides. Double fold top and ends to seal packet, leaving room for heat circulation inside. Repeat to make four packets.
5. Bake for 16 to 20 minutes in the oven or grill 12 to 15 minutes.

Nutritional Analysis
(per serving, excluding unknown items)

263	Calories
9 gm.	Fat
29.1%	Calories from Fat
25 gm.	Protein
23 gm.	Carbohydrate
2 gm.	Dietary Fiber
71 mg.	Cholesterol
241 mg.	Sodium

Exchanges

3	Lean Meat
½	Vegetable
1	Fat
1	Other Carbohydrates

Salmon Oriental in Packets

Serves 4

1 pound salmon fillets
1 package Oriental vegetable mixture
¼ cup soy sauce, low sodium
¼ cup hoisin sauce
¼ teaspoon red pepper flakes
4 sheets of aluminum foil (12x18 inches each)

1. Preheat oven to 450°F or preheat grill to medium-high.
2. Mix soy sauce, hoisin sauce, and red pepper flakes together and set aside.
3. Center one salmon fillet on each sheet of aluminum foil. Top with vegetables. Spoon sauce over salmon and vegetables.
4. Bring up foil sides. Double fold top and ends to seal packet, leaving room for heat circulation inside. Repeat to make four packets.
5. Bake 16 to 20 minutes on a cookie sheet in oven or grill 12 to 15 minutes in covered grill.

Nutritional Analysis
(per serving,
excluding unknown items)

184	Calories
4 gm.	Fat
22.6%	Calories from Fat
25 gm.	Protein
10 gm.	Carbohydrate
1 gm.	Dietary Fiber
59 mg.	Cholesterol
941 mg.	Sodium

Exchanges

3	Lean Meat
½	Vegetable
½	Other Carbohydrates

Sautéed Sea Bass

Serves 4

1½ pounds sea bass fillets
½ cup flour, all-purpose
½ teaspoon sea salt
½ teaspoon black pepper
2 teaspoons olive oil
1 small lemon
2 tablespoons parsley

1. Rinse sea bass and pat dry.
2. Mix salt and pepper with flour on a plate and dip fish to coat all sides.
3. Preheat large skillet with 2 teaspoons of olive oil over medium-high heat until hot, but not smoking. Place fish in pan and sauté until golden, about 3 to 4 minutes. Turn fish over and cook until flesh is no longer translucent, about 3 more minutes, depending on thickness of fish.
4. Garnish with lemon slices and parsley.

Nutritional Analysis
(per serving, excluding unknown items)

246	Calories
6 gm.	Fat
21.9%	Calories from Fat
33 gm.	Protein
14 gm.	Carbohydrate
1 gm.	Dietary Fiber
70 mg.	Cholesterol
353 mg.	Sodium

Exchanges

1	Grain (Starch)
4½	Lean Meat
½	Fat

Scallops with Orzo in Packets

Serves 4

1 pound scallops

3 cups cooked orzo

12 ounces artichoke hearts

2 cups cherry tomatoes, halved

2 cloves garlic, chopped

2 tablespoons lemon juice

1 tablespoon olive oil

¼ teaspoon sea salt

½ teaspoon white pepper

½ teaspoon dill weed

4 ounces feta cheese, crumbled

4 sheets of aluminum foil (12x18 inches each)

1. Preheat oven to 450°F or grill to medium-high.
2. Combine scallops, orzo, artichoke hearts, tomatoes, garlic, lemon juice, olive oil, salt, white pepper, and dill weed.
3. Center one-fourth of mixture on each sheet of aluminum foil.
4. Bring up foil sides. Double fold top and ends to seal packet, leaving room for heat circulation inside. Repeat to make four packets.
5. Bake 14 to 18 minutes on a cookie sheet in oven, or grill 9 to 11 minutes in covered grill. Sprinkle with feta cheese before serving.

Nutritional Analysis (per serving, excluding unknown items)

427	Calories
11 gm.	Fat
23.7%	Calories from Fat
32 gm.	Protein
50 gm.	Carbohydrate
7 gm.	Dietary Fiber
63 mg.	Cholesterol
706 mg.	Sodium

Exchanges

2	Grain (Starch)
3½	Lean Meat
2½	Vegetable
1½	Fat

Scallops in White Wine

Serves 4

Excellent with buttered angel hair pasta or on top of mashed potatoes.

1½ pounds scallops, large
2 cups white wine
4 medium shallots, chopped
2 pounds mushroom, sliced
2 tablespoons parsley
½ teaspoon kosher salt
½ teaspoon white pepper
½ teaspoon marjoram

1. Wash scallops and pat dry.
2. In large nonstick skillet, simmer scallops in wine for five minutes. Remove scallops and set aside.
3. Bring liquid to a boil.
4. Add shallots and mushrooms.
5. Continue boiling until most liquid is absorbed.
6. Reduce heat to simmer. Add scallops, parsley, salt, white pepper, and marjoram. Gently toss to coat scallops. Serve immediately.

Nutritional Analysis (per serving, excluding unknown items)

294	Calories
2 gm.	Fat
8.7%	Calories from Fat
33 gm.	Protein
17 gm.	Carbohydrate
3 gm.	Dietary Fiber
56 mg.	Cholesterol
526 mg.	Sodium

Exchanges

4	Lean Meat
2½	Vegetable

Sea Bass Oriental in Packets

Serves 4

1½ pounds sea bass

16 ounces mixed vegetables, frozen

1 tablespoon hoisin sauce

⅓ cup rice wine vinegar

4 sheets of aluminum foil (12x18 inches each)

1. Preheat oven to 450°F or grill to medium-high.
2. Center one fish fillet on each sheet of aluminum foil. Top with vegetables. Mix hoisin sauce and rice wine vinegar together and spoon over fish and vegetables.
3. Bring up foil sides. Double fold top and ends to seal packet, leaving room for heat circulation inside. Repeat to make four packets.
4. Bake 16 to 20 minutes on a cookie sheet in oven, or grill for 12 to 16 minutes in covered grill.

*Nutritional Analysis
(per serving,
excluding unknown items)*

249	Calories
4 gm.	Fat
14.8%	Calories from Fat
35 gm.	Protein
18 gm.	Carbohydrate
5 gm.	Dietary Fiber
70 mg.	Cholesterol
234 mg.	Sodium

Exchanges

4½	Lean Meat
3	Vegetable

Sea Bass and Vegetables in Packets

Serves 4

1½ pounds sea bass fillets
½ teaspoon dried thyme
½ teaspoon dried marjoram
4 teaspoons lemon juice
2 tablespoons butter
1 pound frozen mixed vegetables
sea salt
white pepper
4 sheets of aluminum foil (12x18 inches each)

1. Preheat oven to 450°F or grill to medium-high.
2. Center one fish fillet on each sheet of aluminum foil. Sprinkle fish with thyme, marjoram, and lemon juice.
3. Place frozen vegetables next to fish on each foil sheet. Sprinkle fish and vegetables with sea salt and white pepper to taste. Dot with butter.
4. Bring up foil sides. Double fold top and ends to seal packet, leaving room for heat circulation inside. Repeat to make four packets.
5. Bake 18 to 22 minutes on a cookie sheet in oven, or grill 16 to 20 minutes in covered grill.

Nutritional Analysis (per serving, excluding unknown items)

291	Calories
10 gm.	Fat
30.0%	Calories from Fat
35 gm.	Protein
16 gm.	Carbohydrate
5 gm.	Dietary Fiber
85 mg.	Cholesterol
228 mg.	Sodium

Exchanges

4½	Lean Meat
3	Vegetable
1	Fat

Main Dishes—Pasta

Baked Mostaccioli with Tomato Basil Sauce
Beef Strips with Spaghetti
Beef Stroganoff
Beef and Tomato Bow Tie Pasta
Blackened Chicken Pasta Salad
Bow Ties with Asparagus and Parmesan Cheese
Italian Sausage Artichoke Pasta
Italian Sausage Pasta Stew
Linguine and Shrimp
Pasta and Grilled Veggies
Pasta with Parmesan Curls and Basil
Pasta with Vodka Sauce
Shrimp and Zucchini with Fettucine
Spinach Noodle Bake
Thai Pasta
Tomato Prosciutto Pasta

Baked Mostaccioli with Tomato Basil Sauce

Serves 6

56 ounces tomatoes, canned, whole

1 tablespoon olive oil

½ teaspoon garlic powder

1 medium onion, chopped

½ teaspoon black pepper

½ teaspoon cayenne pepper

1 pound pasta, mostaccioli

½ cup parsley

3 tablespoons fresh basil

2 ounces Parmesan cheese

1 cup mozzarella cheese, part skim milk, shredded

Nutritional Analysis (per serving, excluding unknown items)

438	Calories
8 gm.	Fat
15.7%	Calories from Fat
22 gm.	Protein
71 gm.	Carbohydrate
5 gm.	Dietary Fiber
18 mg.	Cholesterol
676 mg.	Sodium

Exchanges

4	Grain (Starch)
1½	Lean Meat
2½	Vegetable
½	Fat

1. Preheat oven to 350°F.
2. Cook mostaccioli according to package directions.
3. Place tomatoes in a food processor or blender. Process until smooth.
4. Heat a large Dutch oven, add olive oil.
5. Add onion and sauté until soft.
6. Add tomatoes, black pepper, and cayenne pepper. Simmer 10 minutes uncovered.
7. Add fresh basil, parsley, and Parmesan cheese. Remove from heat. Add cooked mostaccioli to Dutch oven, sprinkle with mozzarella cheese, and bake until cheese is melted.

Beef Strips with Spaghetti

Serves 4

- 12 ounces spaghetti, whole wheat
- 1 pound round steak
- ½ teaspoon black pepper
- ½ teaspoon sea salt
- 1 tablespoon olive oil
- ¼ cup vermouth
- 2 cloves garlic, chopped
- 1 small zucchini
- 1 cup cherry tomatoes, cut in half
- ½ cup Italian salad dressing, low calorie
- 2 tablespoons Parmesan cheese

1. In a large pot, cook spaghetti until al dente.
2. Cut the steak in half lengthwise and then crosswise into 1-inch strips and sprinkle with salt and pepper.
3. Heat olive oil in a large nonstick skillet over medium-high heat. Add half the steak and cook for 1 to 1½ minutes, or until the surface is no longer pink, stirring constantly. Using a slotted spoon, transfer to a plate. Repeat with the remaining beef. Remove from pan and add to already cooked beef. Keep warm.
4. Add the vermouth to the pan to deglaze to get the brown bits off the bottom of the pan. Add garlic and cook for 1 minute, stirring constantly.
5. Add zucchini, tomatoes, and dressing and cook until hot. Add pasta to pan and toss to mix well.
6. Serve in large pasta bowl or individual plates with beef strips on top. Sprinkle with Parmesan cheese.

Nutritional Analysis
(per serving, excluding unknown items)

625	Calories
22 gm.	Fat
31.9%	Calories from Fat
36 gm.	Protein
70 gm.	Carbohydrate
8 gm.	Dietary Fiber
71 mg.	Cholesterol
590 mg.	Sodium

Exchanges

4½	Grain (Starch)
3	Lean Meat
½	Vegetable
2½	Fat

Beef Stroganoff

Serves 4

1 pound beef tenderloin
8 ounces pasta, shells
Olive oil spray
½ teaspoon sea salt
½ teaspoon black pepper
8 ounces mushrooms
½ whole onion, sliced ¼-inch thick
2 teaspoons olive oil
2 tablespoons flour
¾ cup beef broth
½ cup sour cream, light

Nutritional Analysis
(per serving,
excluding unknown items)

606	Calories
30 gm.	Fat
45.0%	Calories from Fat
32 gm.	Protein
51 gm.	Carbohydrate
2 gm.	Dietary Fiber
83 mg.	Cholesterol
543 mg.	Sodium

Exchanges

3	Grain (Starch)
3	Lean Meat
1	Vegetable
4	Fat

1. Cook pasta according to package directions.
2. Meanwhile, trim fat from beef; cut into 1½-inch pieces and sprinkle with salt and pepper.
3. Spray a large nonstick skillet with cooking spray. Heat skillet over medium-high heat until hot. Add beef (half at a time) and stir-fry 1 to 2 minutes or until outside surface is no longer pink. Remove from skillet; keep warm.
4. In same skillet, cook mushrooms and onion in oil for 2 minutes or until tender; stir in flour. Gradually add broth, stirring until blended. Bring to a boil; cook and stir for 2 minutes.
5. Return beef to skillet. Gently stir in sour cream until heated through.
6. Serve beef mixture over pasta.

Beef and Tomato Bow Tie Pasta

Serves 4

One can (28 ounces) whole peeled plum tomatoes, drained and chopped, may be substituted for the fresh tomatoes.

8 ounces pasta, bow tie
1 pound lean ground beef
3 cloves garlic
2 cups tomatoes, chopped
½ teaspoon sea salt
½ teaspoon black pepper
2 tablespoons fresh basil leaves, julienned
3 tablespoons Parmesan cheese

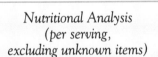

1. In a large pot cook pasta until al dente.
2. In large nonstick skillet, brown ground beef and garlic over medium heat for 8 to 10 minutes or until beef is no longer pink, breaking beef up into ¾-inch crumbles. Drain fat.
3. Stir in tomatoes, salt, and pepper. Cook over medium heat for 5 minutes; stir occasionally.
4. Toss beef mixture and basil with pasta. Sprinkle with Parmesan cheese and serve.

Nutritional Analysis (per serving, excluding unknown items)

340	Calories
25 gm.	Fat
66.8%	Calories from Fat
23 gm.	Protein
5 gm.	Carbohydrate
1 gm.	Dietary Fiber
88 mg.	Cholesterol
392 mg.	Sodium

Exchanges

3	Lean Meat
1	Vegetable
3	Fat

Blackened Chicken Pasta Salad

Serves 6

1 pound boneless, skinless chicken breasts
1 tablespoon Creole seasoning
½ cup grapeseed oil
¼ cup balsamic vinegar
2 tablespoons lemon juice
2 tablespoons Dijon mustard
1 cup apricot nectar
2 teaspoons capers
½ cup scallions
3 medium tomatoes
1 pound pasta, shells
6 cups mixed greens

1. Rub chicken with Creole seasoning. Coat a large, heavy skillet with cooking spray. Add 1 teaspoon of the grapeseed oil and place over medium-high heat until hot. Add chicken and cook 5 to 7 minutes on each side or until chicken is done. Remove chicken from skillet, and let cool. Cut chicken into ½-inch pieces and set aside.
2. Cook pasta in boiling water; drain.
3. To make dressing: Combine remaining grapeseed oil, lemon juice, balsamic vinegar, apricot nectar, and Dijon mustard and mix well.
5. In a large bowl, combine pasta, dressing mixture, tomatoes, scallions, capers, and pasta. Cover and chill thoroughly.
6. Toss mixed greens with pasta mixture. Top with sliced chicken. Serve.

Nutritional Analysis
(per serving, excluding unknown items)

589	Calories
21 gm.	Fat
31.8%	Calories from Fat
30 gm.	Protein
71 gm.	Carbohydrate
5 gm.	Dietary Fiber
44 mg.	Cholesterol
255 mg.	Sodium

Exchanges

4	Grain (Starch)
2½	Lean Meat
1	Vegetable
½	Fruit
3½	Fat

Bow Ties with Asparagus and Parmesan Cheese

Serves 4

20 jumbo asparagus spears
10 ounces pasta, bow tie
1 cup chicken stock
½ cup grated Parmesan cheese
4 tablespoons pine nuts, toasted
Salt and pepper, to taste
4 tablespoons parsley, chopped

1. Bend each asparagus until it breaks, throwing out the end section. Cut the asparagus in ½-inch pieces on an angle.
2. Bring a large pot of water to boil. Put the asparagus pieces in the boiling water for 1 minute to blanch.
3. Remove asparagus and set aside. Add pasta to the boiling water and cook until al dente.
4. In a large saucepan over medium heat, heat pasta in chicken stock. Once pasta mixture has reached a boil, add remaining ingredients and continue to cook stirring constantly until all liquid is gone, about 2 to 3 minutes.
5. Garnish with additional Parmesan cheese if desired.

Nutritional Analysis (per serving, excluding unknown items)

599	Calories
5 gm.	Fat
8.3%	Calories from Fat
24 gm.	Protein
111 gm.	Carbohydrate
5 gm.	Dietary Fiber
8 mg.	Cholesterol
737 mg.	Sodium

Exchanges

7	Grain (Starch)
½	Lean Meat
½	Vegetable

Italian Sausage Artichoke Pasta

Serves 6

1 pound Italian sausage, ground
16 ounces frozen artichoke hearts
2 cloves garlic
2 cups chicken broth
½ cup vermouth, dry, Martini and Rossi®
1 cup sun-dried tomatoes, oil-packed, chopped
1 pound whole wheat pasta, fusilli
½ cup Parmesan cheese
½ cup fresh basil
½ cup parsley
½ teaspoon sea salt
½ teaspoon ground pepper

Nutritional Analysis
(per serving, excluding unknown items)

678	Calories
30 gm.	Fat
40.6%	Calories from Fat
29 gm.	Protein
70 gm.	Carbohydrate
12 gm.	Dietary Fiber
63 mg.	Cholesterol
1201 mg.	Sodium

Exchanges

4	Grain (Starch)
2	Lean Meat
1½	Vegetable
4½	Fat

1. Heat oil in a heavy large skillet over medium-high heat. Add sausage and cook until brown, about 8 minutes. Transfer sausage to a bowl. Add artichokes and garlic to the same skillet and sauté over medium heat until garlic is tender, about 2 minutes. Add broth, vermouth, and sun-dried tomatoes. Boil over medium-high heat until sauce reduces slightly, stirring occasionally, about 8 minutes.

2. Meanwhile, bring a large pot of salted water to a boil. Cook the fusilli in boiling water until al dente, stirring often, about 8 minutes. Drain pasta, but do not rinse. Add pasta, sausage, Parmesan cheese, basil, and parsley to artichoke mixture. Toss until sauce is almost absorbed by the pasta. Season with salt and pepper to taste.

Italian Sausage Pasta Stew

Serves 6

- 1 teaspoon olive oil
- 1 large onion, chopped
- 3 medium carrots, cut ½-inch thick
- 2 cans canned tomatoes, diced
- 1 teaspoon garlic powder
- 1 cup water
- 1 cup pasta, rotini
- 2 medium zucchini cut in slices
- 19¾ ounces cooked Italian sausage links, cut at an angle
- ½ cup Parmesan cheese

1. Heat olive oil in large saucepan. Add onion and sauté 2 to 3 minutes. Add carrots and cook 1 minute. Add tomatoes, water, and rotini; bring to a boil over medium-high heat and cook 5 minutes. Add sausage and zucchini.
2. Continue to cook an additional 5 to 8 minutes, or until pasta and vegetables are tender and stew is slightly thickened. Top each serving with Parmesan cheese.

*Nutritional Analysis
(per serving,
excluding unknown items)*

422	Calories
27 gm.	Fat
58.3%	Calories from Fat
24 gm.	Protein
19 gm.	Carbohydrate
2 gm.	Dietary Fiber
78 mg.	Cholesterol
1055 mg.	Sodium

Exchanges

½	Grain (Starch)
3	Lean Meat
1½	Vegetable
3½	Fat

Linguine and Shrimp

Serves 4

1 pound linguine

1 teaspoon sea salt

3 tablespoons butter

2 tablespoons extra virgin olive oil

4 cloves garlic

1 pound shrimp

¼ teaspoon black pepper

1 tablespoon lemon zest

¼ cup lemon juice

¼ teaspoon red pepper flakes

¼ cup fresh parsley

1. Fill a large pot with water and bring to boil. Add salt and linguine and cook for 7 to 10 minutes, until al dente. Drain pasta, but do not rinse.
2. Meanwhile, in another large (12-inch), heavy-bottomed pan, melt butter and olive oil over medium-low heat. Add garlic and sauté for 1 minute. Add shrimp and pepper. Sauté until shrimp have just turned pink, about five minutes, stirring often. Remove from heat. Add parsley, lemon zest, lemon juice, and red pepper flakes.
3. Add linguine to shrimp mixture and toss to combine. Served immediately.

Nutritional Analysis (per serving, excluding unknown items)

684	Calories
19 gm.	Fat
25.5%	Calories from Fat
38 gm.	Protein
88 gm.	Carbohydrate
3 gm.	Dietary Fiber
196 mg.	Cholesterol
736 mg.	Sodium

Exchanges

5½	Grain (Starch)
3	Lean Meat
3	Fat

Pasta and Grilled Veggies

Serves 4

8 ounces whole wheat macaroni, elbow, cooked and drained
2 whole zucchini, sliced ½-inch thick
1 head cauliflower, cut in 2-inch cubes
1 whole leek, cut in half
1 tablespoon olive oil
1 teaspoon salt
1 tablespoon black pepper
¼ cup rice wine vinegar

1. Preheat grill to medium-high temperature.
2. Toss vegetables in olive oil, salt, and pepper and pour into a vegetable grate pan.
3. Place on grill and cook until slightly charred.
4. Place pasta in large serving bowl; add rice wine vinegar and grilled vegetables and toss.
5. Serve warm or chilled.

Nutritional Analysis (per serving, excluding unknown items)

267	Calories
4 gm.	Fat
13.9%	Calories from Fat
10 gm.	Protein
52 gm.	Carbohydrate
7 gm.	Dietary Fiber
0 mg.	Cholesterol
553 mg.	Sodium

Exchanges

3	Grain (Starch)
1½	Vegetable
1	Fat

Pasta with Parmesan Curls and Basil

Serves 6

1 pound pasta

1 tablespoon olive oil

4 tablespoons toasted pine nuts

1 cup chicken stock

salt and pepper, to taste

1 cup basil, fresh, torn

8 ounces Parmesan cheese, block

1. Cook pasta in a large pot of boiling water, al dente. Drain, but do not rinse.
2. Heat olive oil in a large skillet. Add pine nuts and stir until golden. Add chicken stock and bring to boil. Add pasta and toss to coat. Add salt, pepper, and torn basil. Slice Parmesan cheese in curls, using a vegetable peeler or microplane slicer, on top of pasta.

Nutritional Analysis (per serving, excluding unknown items)

306	Calories
4 gm.	Fat
10.6%	Calories from Fat
10 gm.	Protein
57 gm.	Carbohydrate
2 gm.	Dietary Fiber
0 mg.	Cholesterol
363 mg.	Sodium

Exchanges

3½	Grain (Starch)
½	Fat

Pasta with Vodka Sauce

Serves 4

8 ounces pasta
2 teaspoons olive oil
2 cloves garlic, chopped
½ teaspoon tarragon
14½ ounces diced tomatoes
8 ounces tomato sauce
¼ cup vodka
1 teaspoon cornstarch
½ cup milk, 1% lowfat
¼ cup fresh basil, chopped
¼ cup Parmesan cheese, grated

1. Cook pasta according to package instructions.
2. Heat oil in a large skillet over medium-high heat. Add garlic and sauté 1 minute. Add tarragon and stir to coat. Add diced tomatoes, tomato sauce, and vodka; simmer 10 minutes.
3. Dissolve cornstarch in milk and whisk until blended. Add mixture to skillet and simmer 1 minute, until sauce thickens. Add cooked pasta and basil and toss to coat.
4. Transfer pasta to large pasta bowl and sprinkle fresh Parmesan over top.

Nutritional Analysis
(per serving, excluding unknown items)

343	Calories
5 gm.	Fat
15.6%	Calories from Fat
12 gm.	Protein
54 gm.	Carbohydrate
3 gm.	Dietary Fiber
5 mg.	Cholesterol
465 mg.	Sodium

Exchanges

3	Grain (Starch)
½	Lean Meat
1½	Vegetable
½	Fat

Shrimp and Zucchini with Fettucine

Serves 6

1 pound fettucine
1 cup low sodium chicken broth
4 small zucchini, sliced ⅛-inch thick
3 medium tomatoes, diced
1 pound shrimp, cooked
½ teaspoon garlic powder
¼ cup parsley, chopped
1 tablespoon olive oil
sea salt
white pepper

1. Cook fettucine in a large pot of boiling water until al dente.
2. Add chicken broth to large skillet and bring to boil. Keep boiling until stock is reduced to ½ cup. Reduce heat to simmer.
3. Add zucchini, tomatoes, shrimp, and garlic powder to broth. Stir until all ingredients are heated thoroughly.
4. Add hot cooked fettucine, parsley, and olive oil to pan and toss.
5. Season with salt and pepper to taste. Serve immediately.

---

*Nutritional Analysis
(per serving,
excluding unknown items)*

417	Calories
5 gm.	Fat
10.0%	Calories from Fat
29 gm.	Protein
64 gm.	Carbohydrate
4 gm.	Dietary Fiber
148 mg.	Cholesterol
272 mg.	Sodium

Exchanges

4	Grain (Starch)
2½	Lean Meat
1	Vegetable
½	Fat

Spinach Noodle Bake

Serves 6

1½ cups egg substitute
1 cup cottage cheese, 2% fat, whipped
2 tablespoons dried onion flakes
½ teaspoon nutmeg
¼ teaspoon sea salt
½ teaspoon white pepper
10 ounces frozen spinach
8 ounces egg noodles, wide

1. Preheat oven to 350°F.
2. Cook noodles in a large pot of boiling water, cooking slightly underdone as they will continue cooking in oven.
3. In large bowl, beat together egg substitute, cottage cheese, onion flakes, nutmeg, sea salt, and white pepper until well blended. Gently stir in spinach and noodles.
4. Evenly coat an 11x7x1½-inch baking dish with nonstick spray. Spread noodle mixture in dish.
5. Cover with aluminum foil. Bake until knife inserted near center comes out clean (about 35 to 45 minutes).

Nutritional Analysis
(per serving, excluding unknown items)

292	Calories
9 gm.	Fat
28.3%	Calories from Fat
19 gm.	Protein
34 gm.	Carbohydrate
3 gm.	Dietary Fiber
40 mg.	Cholesterol
394 mg.	Sodium

Exchanges

2	Grain (Starch)
1½	Lean Meat
½	Vegetable
½	Fat

Thai Pasta

Serves 6

1 pound pasta, linguine
1 teaspoon grapeseed oil
4 medium carrots, julienned
½ teaspoon red chili flakes
1 large cucumber, julienned
4 each scallions

SAUCE:

3 tablespoons peanut butter
⅓ cup low sodium soy sauce
¼ cup low sodium chicken broth
¼ teaspoon black pepper

1. Prepare pasta according to package directions; drain and transfer to a serving bowl. Warm 1 teaspoon grapeseed oil in a large wok or nonstick skillet over high heat. Add carrots and stir-fry for 2 to 3 minutes until tender. Add red chili flakes, cucumbers, and scallions to wok or skillet. Stir-fry for 2 minutes.
2. Add all sauce ingredients to wok or skillet. Bring to a boil. Pour vegetables with sauce over pasta. Toss well and serve immediately.

Nutritional Analysis
(per serving,
excluding unknown items)

376	Calories
6 gm.	Fat
14.8%	Calories from Fat
14 gm.	Protein
67 gm.	Carbohydrate
5 gm.	Dietary Fiber
0 mg.	Cholesterol
617 mg.	Sodium

Exchanges

4	Grain (Starch)
½	Lean Meat
1½	Vegetable
1	Fat

Tomato Prosciutto Pasta

Serves 6

Keep prosciutto in your freezer to have on hand.

1 tablespoon olive oil
½ teaspoon garlic paste
4 ounces prosciutto, diced
56 ounces tomatoes, canned
1 cup basil, fresh
3 ounces Parmesan cheese
14 ounces pasta, Bucatini Rigati

1. Cook pasta according to package directions.
2. Using a Dutch oven, heat olive oil. Add garlic and prosciutto; sauté until golden.
3. Add tomatoes and cover, leaving room for steam to escape so sauce will thicken. Let simmer 20 minutes.
4. Pour sauce over hot pasta. Toss with fresh basil and Parmesan cheese.

Nutritional Analysis (per serving, excluding unknown items)

419	Calories
9 gm.	Fat
20.2%	Calories from Fat
22 gm.	Protein
62 gm.	Carbohydrate
5 gm.	Dietary Fiber
24 mg.	Cholesterol
1170 mg.	Sodium

Exchanges

3½	Grain (Starch)
1½	Lean Meat
2	Vegetable
1	Fat

Main Dishes—Chicken

Apricot Roasted Chicken
Asian BBQ Chicken Thighs
Asian Chicken Breasts
Buffalo Chicken
Chicken Breasts with Capers
Chicken Breasts in Sour Cream Sauce
Chicken Noodle Soup
Chicken in Red Wine
Chicken Wraps
Lemon Chicken and Rice
Pineapple Chicken Breast
Roast Chicken

Apricot Roasted Chicken

Serves 4

4 chicken breast halves skinless
4 chicken thighs skinless
6 ounces apricot preserves
8 prunes, pitted and cut in half
¼ cup olive oil
1 tablespoon white wine vinegar
¼ teaspoon sea salt
¼ teaspoon black pepper
½ teaspoon garlic powder
½ teaspoon ground sage

1. Preheat oven to 400°F.
2. Toss all of the ingredients together with the chicken until the chicken is evenly coated with the sauce. Arrange the chicken pieces in the pan, spaced evenly apart.
3. Roast until the chicken is thoroughly cooked, and the juices run clear, about 30 to 40 minutes.

Nutritional Analysis
(per serving, excluding unknown items)

477	Calories
18 gm.	Fat
33.4%	Calories from Fat
42 gm.	Protein
39 gm.	Carbohydrate
2 gm.	Dietary Fiber
126 mg.	Cholesterol
271 mg.	Sodium

Exchanges

5½	Lean Meat
½	Fruit
2½	Fat
2	Other Carbohydrates

Asian BBQ Chicken Thighs

Serves 4

1 cup soy sauce, low sodium
2 tablespoons hoisin sauce
½ teaspoon five-spice powder
8 chicken thighs skinless

1. Mix soy sauce, hoisin sauce, and five-spice powder in a large plastic bag.
2. Place chicken thighs in bag, coating all sides of chicken, seal, and refrigerate for up to 24 hours and as little as 15 minutes.
3. Grill on medium heat until thoroughly cooked.

Nutritional Analysis
(per serving,
excluding unknown items)

220	Calories
6 gm.	Fat
23.9%	Calories from Fat
31 gm.	Protein
10 gm.	Carbohydrate
1 gm.	Dietary Fiber
115 mg.	Cholesterol
2648 mg.	Sodium

Exchanges

4	Lean Meat
1½	Vegetable

Asian Chicken Breasts

Serves 4

4 medium chicken breast halves skinless
4 tablespoons hoisin sauce
2 teaspoons scallions, chopped

1. Preheat oven to 375°F.
2. Spread 1 tablespoon hoisin sauce on each chicken breast.
3. Roast in oven for 20 minutes or until chicken reaches 170°F internal temperature.
4. Garnish with chopped scallions and serve.

Nutritional Analysis (per serving, excluding unknown items)

165	Calories
2 gm.	Fat
11.4%	Calories from Fat
28 gm.	Protein
7 gm.	Carbohydrate
Trace	Dietary Fiber
69 mg.	Cholesterol
335 mg.	Sodium

Exchanges

4	Lean Meat
1/2	Other Carbohydrates

Buffalo Chicken

Serves 4
---

2 teaspoons olive oil

1½ pounds chicken tenders

½ teaspoon kosher salt

½ teaspoon black pepper

½ teaspoon red chili flakes

½ teaspoon garlic powder

½ teaspoon paprika

1 cup light sour cream

2 ounces blue cheese, crumbled

4 stalks celery, sliced

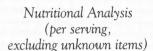

1. Coat a large nonstick skillet with olive oil and heat over medium-high heat.
2. Place chicken on a plate and sprinkle with salt, pepper, red chili flakes, garlic powder, and paprika. Add chicken to skillet and cook, turning occasionally until lightly browned and cooked through, about 5 to 7 minutes.
3. Mix light sour cream and crumbled blue cheese together and serve chicken with blue cheese dressing and celery on the side.

Nutritional Analysis
(per serving,
excluding unknown items)

268	Calories
9 gm.	Fat
29.2%	Calories from Fat
42 gm.	Protein
6 gm.	Carbohydrate
1 gm.	Dietary Fiber
100 mg.	Cholesterol
781 mg.	Sodium

Exchanges

½	Lean Meat
1	Fat

Chicken Breasts with Capers

Serves 4

1 pound boneless chicken breasts
3 tablespoons flour
¾ teaspoon black pepper
¾ teaspoon lemon peel
½ cup dry vermouth, Martini & Rossi®
½ cup low sodium chicken broth
¼ cup lemon juice
4 lemon slices
¼ cup capers

1. Dredge chicken breasts lightly in flour and lemon pepper. Melt butter in large skillet over medium-high heat. Sauté breasts, turning once, until golden brown, about 5 to 6 minutes.

2. Add vermouth and chicken broth to skillet; shake pan gently, reduce heat to medium, cover, and cook for 5 minutes. Remove chicken to a warm serving plate, add the lemon juice and capers to the pan, bring the heat up to medium-high, and cook until the sauce is slightly thickened. Spoon sauce over the chicken breast. Garnish with lemon slices.

---

*Nutritional Analysis
(per serving,
excluding unknown items)*

209	Calories
2 gm.	Fat
10.6%	Calories from Fat
29 gm.	Protein
9 gm.	Carbohydrate
Trace	Dietary Fiber
66 mg.	Cholesterol
228 mg.	Sodium

Exchanges

½	Grain (Starch)
4	Lean Meat

Chicken Breasts in Sour Cream Sauce

Serves 4

¼ cup flour
½ teaspoon kosher salt
½ teaspoon white pepper
2 teaspoons grapeseed oil
1 tablespoon butter
4 medium chicken breast halves skinless
½ cup dry vermouth
1 cup chicken broth
1 cup light sour cream
¼ cup black olives, sliced

1. Mix flour, salt, and pepper together and coat chicken breasts.
2. Heat grapeseed oil and butter in a large skillet and brown chicken on both sides. Pour in dry vermouth, cover, and simmer for 15 minutes.
3. Remove chicken from pan and keep warm. Add remaining flour mixture into chicken broth and pour into pan. Bring to boil, continually stirring to reduce liquid until thickened, about 2 minutes.
4. Reduce heat and stir sour cream into pan. Spoon sauce over chicken and garnish with black olives.

Nutritional Analysis (per serving, excluding unknown items)

279	Calories
9 gm.	Fat
33.0%	Calories from Fat
31 gm.	Protein
11 gm.	Carbohydrate
1 gm.	Dietary Fiber
81 mg.	Cholesterol
625 mg.	Sodium

Exchanges

½	Grain (Starch)
4	Lean Meat
1½	Fat

Chicken Noodle Soup

Serves 6

2 quarts low sodium chicken broth

1½ pounds chicken tenders, cut in 1-inch pieces

1 medium onion, chopped

3 whole carrots, peeled and chopped

3 stalks celery, sliced

2 tablespoons butter

1 cup mushrooms, quartered

1 tablespoon fresh lemon juice

8 ounces egg noodles, dried

½ cup parsley, chopped

Sea salt and black pepper to taste

1. Combine chicken broth, chicken, onions, carrots, and celery in heavy large pot. Bring to a boil. Reduce heat, cover partially for 10 minutes.
2. Melt 2 tablespoons butter in heavy large skillet over medium-high heat.
3. Add mushrooms and sauté until they begin to brown, about 5 minutes. Stir in lemon juice. Add mushrooms to chicken broth mixture; stir in noodles.
4. Simmer until noodles are tender, about 5 minutes. Season soup to taste with salt and pepper.

Nutritional Analysis
(per serving,
excluding unknown items)

388	Calories
6 gm.	Fat
14.9%	Calories from Fat
46 gm.	Protein
37 gm.	Carbohydrate
3 gm.	Dietary Fiber
103 mg.	Cholesterol
971 mg.	Sodium

Exchanges

2	Grain (Starch)
1½	Lean Meat
1½	Vegetable
1	Fat

Chicken in Red Wine

___

Serves 6
___

This is a one-dish meal. Serve with a crusty bread and tossed salad.

6 each chicken breast halves, skinless
12 frozen pearl onions, thawed
12 small new potatoes
6 small carrots
4 stalks celery
1 teaspoon garlic powder
1 teaspoon dried rosemary
1 teaspoon dried thyme
1 teaspoon black pepper
2 teaspoons olive oil
1½ cups low sodium chicken broth
½ cup tomato paste
1 pound mushrooms
2 cups red wine
4 tablespoons flour, bleached

Nutritional Analysis
(per serving, excluding unknown items)

546	Calories
4 gm.	Fat
7.4%	Calories from Fat
41 gm.	Protein
77 gm.	Carbohydrate
11 gm.	Dietary Fiber
68 mg.	Cholesterol
896 mg.	Sodium

Exchanges

3	Grain (Starch)
4	Lean Meat
5	Vegetable
½	Fat

1. Shake chicken in a bag with flour, garlic, rosemary, thyme, and pepper.
2. In a large Dutch oven, heat olive oil and brown chicken on both sides. Remove from pan. Add pearl onions to pan and brown. Add potatoes, carrots, and celery and place chicken back in pan. Pour the red wine in to deglaze the pan.
3. Mix tomato paste with chicken broth along with leftover flour mixture and pour over chicken and vegetables.
4. Add mushrooms, cover, and reduce heat to a simmer. Cook for 45 minutes.

(continued on next page)

5. Remove chicken breasts and set inside. Bring liquid to a boil, continually stirring until reduced and thickened, about 2 minutes.
6. Reduce heat to low, return chicken to mixture, and warm thoroughly.

Chicken Wraps

Serves 4

- 2 tablespoons grapeseed oil
- 1½ pounds chicken tenders, chopped
- 2 cups sliced mushrooms
- ½ teaspoon sea salt
- ½ teaspoon black pepper
- 3 cloves garlic, chopped
- 1 tablespoon ginger root, grated
- 1 medium orange
- 1 whole red bell pepper, chopped
- 8 ounces canned water chestnuts, chopped
- 2 whole scallions, chopped
- 3 tablespoons hoisin sauce
- 1 head Boston lettuce, washed and patted dry
- 1 orange, cut in wedges for garnish

Nutritional Analysis (per serving, excluding unknown items)

329	Calories
9 gm.	Fat
23.7%	Calories from Fat
41 gm.	Protein
24 gm.	Carbohydrate
4 gm.	Dietary Fiber
85 mg.	Cholesterol
738 mg.	Sodium

Exchanges

2½	Vegetable
1½	Fat
½	Other Carbohydrates

1. Heat grapeseed oil in a large skillet or wok, add chopped chicken tenders into pan and stir fry for 1 to 2 minutes. Add mushrooms and continue cooking another 2 minutes.
2. Add salt, pepper, garlic, and ginger; cook for 1 minute, stirring constantly.
3. Zest the orange into the chicken mixture. Add red bell pepper and water chestnuts and cook for one minute, stirring constantly.
4. Add hoisin sauce and toss to coat the mixture evenly.
5. Transfer the chicken mixture to serving platter and pile the lettuce leaves on the side. Add wedges of orange to platter to garnish. To eat, pile spoonfuls into lettuce leaves, wrapping lettuce around filling.

Lemon Chicken and Rice

Serves 4

1 tablespoon olive oil
1 pound chicken tenders
1 clove garlic
1 cup brown rice
2½ cups chicken broth
¼ cup lemon juice
1 tablespoon lemon zest
½ teaspoon black pepper

1. Heat olive oil in large skillet over medium-high heat until hot.
2. Add chicken and garlic; sauté, continually stirring until browned.
3. Stir in rice, lemon juice, and broth. Reduce heat to simmer.
4. Cover and cook 30 minutes or until liquid is absorbed.
5. Stir in lemon peel and pepper. Serve immediately.

Nutritional Analysis (per serving, excluding unknown items)

345	Calories
6 gm.	Fat
16.7%	Calories from Fat
32 gm.	Protein
40 gm.	Carbohydrate
1 gm.	Dietary Fiber
56 mg.	Cholesterol
679 mg.	Sodium

Exchanges

2½	Grain (Starch)
½	Lean Meat
1	Fat

Pineapple Chicken Breast

Serves 6

6 boneless skinless chicken breasts
12 ounces teriyaki marinade
20 ounces canned pineapple slices
1 small red onion, sliced ¼-inch thick
1 medium green bell pepper, sliced ½-inch thick

1. Drain pineapple liquid from can and reserve.
2. Deeply pierce chicken in several places with fork. In large resealable plastic bag, combine chicken, ⅔ cup teriyaki marinade, and ¼ cup of pineapple liquid; seal and marinate in refrigerator 30 minutes, turning occasionally.
3. Preheat oven to 375°F.
4. Spray 13x9x2-inch baking dish with nonstick vegetable spray; add red onion and green pepper.
5. Remove chicken from bag, discarding used marinade.
6. Place chicken over vegetables and top with pineapple slices. Drizzle remaining marinade over top.
7. Bake in oven until chicken reaches 175° to 185°F, about 40 to 45 minutes.

Nutritional Analysis
(per serving, excluding unknown items)

372	Calories
3 gm.	Fat
7.7%	Calories from Fat
55 gm.	Protein
28 gm.	Carbohydrate
1 gm.	Dietary Fiber
137 mg.	Cholesterol
2123 mg.	Sodium

Exchanges

7½	Lean Meat
½	Vegetable
½	Fruit
1	Other Carbohydrates

Roast Chicken

Serves 4

4 chicken legs, skinless
4 chicken breast halves, skinless
1 large yellow onion, sliced 1-inch thick
4 stalks celery, sliced 1-inch thick
4 medium carrots, sliced 1-inch thick
1 tablespoon olive oil
½ teaspoon kosher salt
½ teaspoon black pepper
2 medium potatoes, cubed

1. Preheat oven to 425°F.
2. Toss all ingredients together and spread out in a roasting pan.
3. Roast in oven for approximately 35 minutes or until chicken reaches 170°F.
4. Let stand 10 minutes before serving.

Nutritional Analysis
(per serving,
excluding unknown items)

412   Calories
10 gm.   Fat
22.4%   Calories from Fat
56 gm.   Protein
22 gm.   Carbohydrate
4 gm.   Dietary Fiber
172 mg. Cholesterol
488 mg. Sodium

Exchanges

½   Grain (Starch)
7½   Lean Meat
2   Vegetable
½   Fat

Main Dishes—Eggs

Asparagus Quiche
Baked Eggs
French Toast
Ham and Cheese Quiche
Omelet

Asparagus Quiche

Serves 6

½ cup mozzarella cheese, part skim milk

¼ cup cheddar cheese

2 cups asparagus

¼ cup leeks, thinly sliced

½ cup mushrooms, sliced ¼-inch thick

3 cups egg substitute, liquid

2 cups milk, 1% lowfat

½ teaspoon sea salt

½ teaspoon black pepper

1. Preheat oven to 375°F.
2. Bend each asparagus until it breaks, throwing out the end section. Cut the asparagus in ½-inch pieces on the angle.
3. Grease a 10x12-inch pan. Spread mozzarella and cheddar cheese on bottom of pan and top with vegetables.
4. Whisk together egg substitute, milk, salt, and pepper. Pour over cheese and vegetables. Bake for about 45 minutes or until knife inserted in center comes out clean.

---

*Nutritional Analysis
(per serving,
excluding unknown items)*

199	Calories
8 gm.	Fat
37.9%	Calories from Fat
23 gm.	Protein
8 gm.	Carbohydrate
1 gm.	Dietary Fiber
15 mg.	Cholesterol
501 mg.	Sodium

Exchanges

2½	Lean Meat
½	Vegetable
½	Non-fat Milk
½	Fat

Baked Eggs

___
Serves 2
___

Serve hot with toasted bread.

¼ teaspoon fresh thyme
¼ teaspoon fresh rosemary
1 tablespoon fresh parsley
1 tablespoon Parmesan cheese
4 large eggs
2 tablespoons butter
½ teaspoon sea salt
½ teaspoon black pepper

1. Place the oven rack 6 inches below the broiler. Preheat oven to broil setting for approximately 5 minutes.
2. Combine the thyme, rosemary, parsley, and Parmesan cheese and set aside.
3. Place two individual gratin dishes on a baking sheet. Place 1 tablespoon of butter in each dish and place under the broiler for about three minutes, until hot and bubbly. Remove from oven.
4. Crack two eggs into a teacup and pour eggs into each gratin dish. Sprinkle evenly with the herb mixture; then sprinkle liberally with salt and pepper.
5. Place back under the broiler for 5 to 6 minutes until the whites of the eggs are almost cooked. The eggs will continue to cook after you take them out of the oven. Allow to set for 60 seconds prior to serving.

Nutritional Analysis (per serving, excluding unknown items)

263	Calories
22 gm.	Fat
76.6%	Calories from Fat
14 gm.	Protein
1 gm.	Carbohydrate
Trace	Dietary Fiber
457 mg.	Cholesterol
775 mg.	Sodium

Exchanges

1½	Lean Meat
3	Fat

French Toast

Serves 4

Top with melted butter and maple syrup or vanilla yogurt and fresh berries.

4 whole eggs
1 teaspoon cinnamon
¼ cup milk, 1% lowfat
1 teaspoon vanilla extract
4 slices raisin bread, extra thick
½ teaspoon nutmeg

1. Mix all ingredients, except bread, together in a flat bowl that will hold one slice of bread.
2. Dip each piece of bread in egg mixture, making sure the piece of bread is soaked.
3. Brown bread in large skillet sprayed with nonstick coating.

Nutritional Analysis
(per serving, excluding unknown items)

158	Calories
6 gm.	Fat
37.0%	Calories from Fat
9 gm.	Protein
16 gm.	Carbohydrate
1 gm.	Dietary Fiber
213 mg.	Cholesterol
179 mg.	Sodium

Exchanges

1	Grain (Starch)
1	Lean Meat
½	Fat

Ham and Cheese Quiche

Serves 6

1 whole Pillsbury® pie crust
½ cup green bell pepper, chopped
½ cup onion, chopped
½ cup lean ham, chopped
½ cup cheddar cheese, lowfat, shredded
1½ cups egg substitute
1 cup milk, 1% lowfat
¼ teaspoon paprika
½ teaspoon sea salt
¼ teaspoon white pepper

1. Preheat oven to 425°F.
2. Press the pie crust into a 9-inch quiche pan and bake for 10 minutes.
3. Remove from oven and reduce oven temperature to 375°F.
4. In small skillet over medium heat, sauté bell pepper and onion in butter until tender, but not brown; about 6 minutes.
5. Sprinkle vegetables, ham, and cheese into pie shell.
6. Beat together egg substitute, paprika, milk, salt, and pepper and until well blended. Pour over vegetable and ham mixture.
7. Bake in preheated 375°F oven for 15 minutes. Remove quiche from oven and cover outer rim of crust with foil to avoid burning. Place quiche back in oven and continue cooking until knife inserted near center comes out clean, about 15 to 20 minutes longer. Let stand 10 minutes before serving.

---

*Nutritional Analysis
(per serving,
excluding unknown items)*

313	Calories
18 gm.	Fat
51.3%	Calories from Fat
14 gm.	Protein
24 gm.	Carbohydrate
1 gm.	Dietary Fiber
17 mg.	Cholesterol
649 mg.	Sodium

Exchanges

1	Grain (Starch)
1½	Lean Meat
½	Vegetable
2½	Fat

Omelet

___
Serves 1
___

A simple omelet can sometimes be just the thing. The secret is to cook eggs slowly so they stay tender. You can fill with fresh herbs, vegetables, meats, cheese, or simply enjoy plain.

1 whole egg
2 egg whites
2 tablespoons water
⅛ teaspoon sea salt
⅛ teaspoon white pepper
1 teaspoon butter

1. Whip all ingredients except butter, together.
2. In a 7- to 10-inch omelet pan or skillet over medium-high heat, melt butter until just hot enough to make a drop of water sizzle. Pour in egg mixture. Mixture should set immediately at edges.
3. With inverted pancake turner, carefully push cooked portions at edges toward center so uncooked portions can reach hot pan surface. Tilt pan and move cooked portions, as necessary.
4. When top is thickened and no visible liquid egg remains (this is the point you would add any filling), using a pancake turner, fold omelet in half, or roll. Invert onto plate with a quick flip of the wrist or slide from pan onto plate.

___

Nutritional Analysis
(per serving, excluding unknown items)

142	Calories
9 gm.	Fat
57.5%	Calories from Fat
13 gm.	Protein
1 gm.	Carbohydrate
Trace	Dietary Fiber
222 mg.	Cholesterol
454 mg.	Sodium

Exchanges

1½	Lean Meat
1	Fat

Main Dishes—Pork

Bratwurst Kabobs
Cassoulet
Creole Sausage and Brown Rice
Italian Sausage and Peppers
Paella
Peaches and Pork
Pork with Apples
Pork with Caramelized Onion and Apricots
Pork Chops Caribbean
Pork Chops with Mushrooms
Pork Piccata
Slow-Cooked Pork Stew

Bratwurst Kabobs

Serves 6

½ cup soy sauce, low sodium

¼ cup apple juice, frozen concentrate

3 tablespoons mustard

16 ounces bratwurst links, Johnsonville® Smoked, cut in 1-inch cubes

2 medium red bell peppers, cut in 1-inch cubes

1 medium green bell pepper, cut in 1-inch cubes

1 medium yellow squash, cut in 1-inch cubes

1 medium onion, cut in 1-inch cubes

1. In a resealable plastic bag or bowl, combine soy sauce, apple juice, and mustard; add vegetables. Seal bag or cover container; toss to coat well.
2. Refrigerate for 1 hour.
3. Drain and reserve marinade. Thread sausage cubes and vegetables alternately on skewers. Brush with reserved marinade.
4. Grill over medium-hot coals; turn and baste often, for 15–20 minutes or until sausage is hot and vegetables reach desired doneness. Serve immediately.

Nutritional Analysis (per serving, excluding unknown items)

65	Calories
1 gm.	Fat
7.5%	Calories from Fat
3 gm.	Protein
14 gm.	Carbohydrate
2 gm.	Dietary Fiber
0 mg.	Cholesterol
899 mg.	Sodium

Exchanges

1½	Vegetable
½	Fruit

Cassoulet

Serves 4

Olive oil spray
1 medium onion, chopped
1 clove garlic, chopped
8 ounces Polish sausage link, Johnsonville®, cooked and cut in 2-inch cubes
¼ teaspoon thyme
¾ cup low sodium chicken broth
30 ounces navy beans, canned, rinsed, and drained
2 tablespoons tomato paste
½ cup bread crumbs

1. Preheat broiler.
2. Coat a nonstick ovenproof skillet with cooking spray and heat. Add onion, garlic, and Polish sausage and sauté until onion is tender, about 3 to 4 minutes. Add thyme, broth, beans, and tomato paste; simmer for 10 minutes.
3. Sprinkle bread crumbs over beans. Place skillet under broiler until bread crumbs are browned, about 30 seconds.

Nutritional Analysis (per serving, excluding unknown items)

506	Calories
18 gm.	Fat
31.7%	Calories from Fat
28 gm.	Protein
59 gm.	Carbohydrate
12 gm.	Dietary Fiber
40 mg.	Cholesterol
1728 mg.	Sodium

Exchanges

3½	Grain (Starch)
2	Lean Meat
1	Vegetable
2½	Fat

Creole Sausage and Brown Rice

Serves 6

19¾ ounces bratwurst, Johnsonville® Hot 'n Spicy

1 teaspoon olive oil

1 large onion, chopped

28 ounces canned tomatoes, drained and chopped

2½ cups red and green bell peppers, chopped

½ teaspoon thyme

3 cups brown rice, cooked

1. Cook bratwurst on the grill.
2. Slice cooked bratwurst on the bias.
3. Heat oil in large saucepan. Add onion and cook 2 to 3 minutes.
4. Add tomatoes and cook 4 to 5 minutes or until tomatoes begin to soften.
5. Add peppers, sausage, and thyme; cook 3 to 4 minutes or until peppers are crisp-tender. Serve over rice.

Nutritional Analysis
(per serving, excluding unknown items)

446	Calories
26 gm.	Fat
52.2%	Calories from Fat
17 gm.	Protein
36 gm.	Carbohydrate
4 gm.	Dietary Fiber
56 mg.	Cholesterol
719 mg.	Sodium

Exchanges

1½	Grain (Starch)
2	Lean Meat
2	Vegetable
4	Fat

Italian Sausage and Peppers

Serves 6

19¾ ounces Italian sausage links, Johnsonville® mild or hot
Olive oil spray
2 whole red bell peppers, cut in half
1 whole green bell pepper, cut in half
1 medium onion, cut in half
2 cups marinara sauce
½ pound part skim milk mozzarella cheese

1. Preheat the grill to medium-high heat.
2. Spray bell peppers and onion with olive oil spray and place on a grilling vegetable grate. Cook Italian sausage and vegetables on the grill until done. Remove to a carving board. Cut the sausage link slices on an angle and bell peppers and onion in strips.
3. Preheat oven to 350°F.
4. Divide mixture into six individual baking dishes or in one large baking dish. Heat marinara sauce and spoon over sausage slices. Top with mozzarella slices. Bake in oven until cheese has melted, about 10 minutes.

Nutritional Analysis
(per serving,
excluding unknown items)

499	Calories
38 gm.	Fat
68.1%	Calories from Fat
26 gm.	Protein
14 gm.	Carbohydrate
3 gm.	Dietary Fiber
91 mg.	Cholesterol
1227 mg.	Sodium

Exchanges

½	Grain (Starch)
3½	Lean Meat
1	Vegetable
5½	Fat

Paella

Serves 8

3 tablespoons olive oil
3 cloves garlic, crushed
1 teaspoon red pepper flakes
2 cups arborio rice
¼ teaspoon saffron threads
1 whole bay leaf
1 quart chicken broth
4 sprigs thyme
1½ pounds chicken tenders
1 whole red bell pepper
1 medium onion
1 pound cooked Italian sausage
1 pound shrimp
1 cup frozen peas
¼ cup parsley
1 medium lemon

Nutritional Analysis (per serving, excluding unknown items)

592	Calories
22 gm.	Fat
34.4%	Calories from Fat
49 gm.	Protein
46 gm.	Carbohydrate
2 gm.	Dietary Fiber
173 mg.	Cholesterol
1170 mg.	Sodium

Exchanges

2½	Grain (Starch)
3½	Lean Meat
½	Vegetable
3	Fat

1. Preheat a very wide or paella pan over medium high heat. Add 2 tablespoons olive oil, crushed garlic, red pepper flakes, and rice. Sauté for 2 or 3 minutes.
2. Add saffron threads, bay leaf, broth, and thyme.
3. Bring liquid to a boil over high heat. Cover the pan with lid or foil, reduce heat and simmer for 15 minutes.
4. Season chicken with salt and pepper. Preheat 1 tablespoon olive oil in a separate nonstick skillet, over medium high-heat. Brown chicken on both sides.

5. Add peppers and onions to pan and cook for three minutes.
6. Add Italian sausage to the pan and cook an additional two minutes. Remove pan from heat.
7. After rice has simmered for 15 minutes, remove lid or foil from pan. Add shrimp, nesting them in the cooking rice.
8. Pour in peas, scatter lemon zest over rice and seafood, and cover pan again.
9. After five minutes, remove lid or foil from paella. Stir rice and seafood mixture, lifting out bay leaf and thyme stems.
10. Arrange cooked chicken, peppers, onions, and Italian sausage around the pan.
11. Top with parsley. Serve with wedges of lemon.

Peaches and Pork

Serves 6

3 tablespoons Southern Comfort®
1½ pounds pork shoulder, cut in 2-inch cubes
4 peaches, firm, cut in 2-inch chunks
1 teaspoon dried thyme
2 tablespoons olive oil
½ teaspoon sea salt
½ teaspoon black pepper

1. Preheat grill to medium temperature.
2. Combine pork, peaches, onion, thyme, bay leaf, Southern Comfort®, and oil in a large bowl. Cover and refrigerate, marinating for 1 hour.
3. Skewer pork, peaches, and onion alternately.
4. Grill 5 to 10 minutes, basting occasionally with marinade.

Nutritional Analysis (per serving, excluding unknown items)

294	Calories
20 gm.	Fat
65.6%	Calories from Fat
15 gm.	Protein
8 gm.	Carbohydrate
1 gm.	Dietary Fiber
60 mg.	Cholesterol
212 mg.	Sodium

Exchanges

2	Lean Meat
½	Fruit
3	Fat

Pork with Apples

Serves 6

1½ pounds boneless pork top loin, cut in 6 slices
2 teaspoons butter
½ teaspoons sea salt
½ teaspoons black pepper
3 medium apples, unpeeled
1 teaspoon brown sugar
½ teaspoon cinnamon

1. Season pork with salt and pepper.
2. Melt butter in the large skillet and cook pork loin slices on medium heat, about 5 minutes on each side.
3. Remove to a warm platter. Keep warm.
4. Slice apples ½-inch thick, removing core, and add to pan, along with sugar and cinnamon.
5. Cook over medium heat approximately 10 minutes, turning once or twice until apples begin to brown.
6. Arrange apples around pork and serve.

Nutritional Analysis
(per serving,
excluding unknown items)

263	Calories
18 gm.	Fat
61.1%	Calories from Fat
14 gm.	Protein
11 gm.	Carbohydrate
2 gm.	Dietary Fiber
60 mg.	Cholesterol
53 mg.	Sodium

Exchanges

2	Lean Meat
½	Fruit
2	Fat

Pork with Caramelized Onion and Apricots

Serves 4

1 pound pork tenderloin, cut in 4 pieces
½ teaspoon kosher salt
½ teaspoon black pepper
2 tablespoon butter
1 medium onion, sliced
2 teaspoons sugar
⅓ cup dry vermouth, Martini and Rossi®
2 tablespoons sweet vermouth, Martini and Rossi®
8 each dried apricot halves, cut in thin strips
2 tablespoons sage, cut in thin strips

1. Preheat oven to 350°F.
2. Rinse and dry pork. Salt and pepper both sides.
3. Melt butter in a skillet and brown pork on each side.
4. Remove to an ovenproof dish and bake covered for 30 minutes.
5. In the pan that you browned the pork, add onions and stir. Reduce heat to simmer, cover, and cook for 10 minutes.
6. Uncover the pan and sprinkle the sugar over the cooking onions, stirring constantly until onions have caramelized and are a golden brown in color.
7. Add the dry and sweet vermouth and the apricot strips. Raise the heat and bring to a boil.
8. Continue stirring the mixture until there is no liquid remaining in the pan.

Nutritional Analysis
(per serving, excluding unknown items)

267	Calories
10 gm.	Fat
38.3%	Calories from Fat
25 gm.	Protein
11 gm.	Carbohydrate
2 gm.	Dietary Fiber
89 mg.	Cholesterol
360 mg.	Sodium

Exchanges

3½	Lean Meat
½	Vegetable
½	Fruit
1	Fat

9. Remove the pork from the oven. (Cover pork chops and let rest for 5 minutes.)
10. Pour the juices from the pan used in cooking the pork into the onion mixture. Stir over medium heat.
11. Add sage and mix in.
12. Place pork chops on a large serving platter, or place one pork chop each on four individual dinner plates and top each piece with $\frac{1}{4}$ of the onion mixture.

Pork Chops Caribbean

Serves 4

4 pork chops

¾ cup lime juice

½ teaspoon cayenne pepper

1 tablespoon olive oil

1 large red onion, sliced

1. Combine lime juice and pepper. Marinate pork in mixture and refrigerate 2 to 24 hours.
2. Preheat grill to medium temperature.
3. Remove chops from marinade, reserving marinade. Grill chops for 15 to 18 minutes, turning once.
4. While pork is cooking, heat olive oil in small skillet and sauté onion until soft. Add lime juice marinade and bring to boil. Serve over pork.

---

*Nutritional Analysis
(per serving,
excluding unknown items)*

211	Calories
12 gm.	Fat
49.2%	Calories from Fat
19 gm.	Protein
8 gm.	Carbohydrate
1 gm.	Dietary Fiber
41 mg.	Cholesterol
43 mg.	Sodium

Exchanges

2½	Lean Meat
½	Vegetable
½	Fruit
½	Fat

Pork Chops with Mushrooms

_____

Serves 4

_____

4 pork chops

2 tablespoons flour

½ teaspoon sea salt

¼ teaspoon white pepper

1 teaspoon olive oil

1 teaspoon butter

½ cup dry vermouth

1 cup sliced mushrooms

½ cup light sour cream

1. Combine flour, salt, and pepper. Lightly dredge chops in flour mixture.
2. Heat olive oil and butter in nonstick skillet; add chops and sauté about two minutes on each side.
3. Add vermouth; cook over medium heat about 4 to 5 minutes.
4. Remove chops, set aside and keep warm.
5. Add mushrooms and cook until soft.
6. Add sour cream to skillet; cook over low heat, stirring constantly, just until thickened.
7. Return chops to pan, cook just to reheat, and serve immediately.

Nutritional Analysis (per serving, excluding unknown items)

324	Calories
17 gm.	Fat
55.8%	Calories from Fat
24 gm.	Protein
7 gm.	Carbohydrate
Trace	Dietary Fiber
79 mg.	Cholesterol
319 mg.	Sodium

Exchanges

3½	Lean Meat
1½	Fat

Pork Piccata

Serves 4

1 pound pork loin, cut in 4 cutlets
3 tablespoon flour
¾ teaspoon black pepper
¾ teaspoon lemon peel
2 tablespoons butter
¼ cup lemon juice
½ cup dry vermouth, Martini and Rossi®
4 slices lemon
¼ cup capers

1. Pound cutlets thin (about ⅛-inch thick) evenly.
2. Dredge lightly in flour, black pepper, and lemon peel.
3. Melt butter in large skillet over medium-high heat.
4. Sauté cutlets, turning once, until golden brown, about 5 to 6 minutes (total time).
5. Add vermouth and lemon juice to skillet; shake pan gently and cook 2 to 3 minutes, until sauce is slightly thickened.
6. Serve cutlets garnished with lemon slices and capers.

Nutritional Analysis
(per serving,
excluding unknown items)

219	Calories
10 gm.	Fat
48.9%	Calories from Fat
15 gm.	Protein
8 gm.	Carbohydrate
Trace	Dietary Fiber
51 mg.	Cholesterol
174 mg.	Sodium

Exchanges

½	Grain (Starch)
2	Lean Meat
1	Fat

Slow-Cooked Pork Stew

Serves 4

3 teaspoons olive oil
2 tablespoons flour, all-purpose
½ teaspoon salt
½ teaspoon black pepper
1 pound pork center loin, cut in 2-inch cubes
20 baby carrots
1 large onion
1 teaspoon garlic powder
⅛ teaspoon ground sage
½ teaspoon oregano
½ cup beef broth
14½ ounces canned tomatoes, diced

1. Heat 2 teaspoons olive oil in skillet.
2. Combine flour, salt, pepper, and pork cubes in a bag; shake to coat pork. Brown pork in skillet.
3. Place carrots in a 4- to 5-quart slow cooker. Place pork over carrots.
4. Place 1 teaspoon of olive oil in skillet. Add onion and sauté over medium heat for three minutes. Add onions, garlic powder, sage, and oregano to slow cooker.
5. Pour beef broth into skillet and scrape up any browned bits; add broth and undrained tomatoes to slow cooker. Cover and cook on low heat for 7 to 8 hours.

Nutritional Analysis (per serving, excluding unknown items)

272	Calories
15 gm.	Fat
47.8%	Calories from Fat
20 gm.	Protein
15 gm.	Carbohydrate
3 gm.	Dietary Fiber
56 mg.	Cholesterol
648 mg.	Sodium

Exchanges

2½	Lean Meat
2	Vegetable
1½	Fat

Main Dishes—Meat

Asian Beef Stir-Fry
Beef and Spinach
Beef and Spinach Burritos
Curried Lamb Chops
Flat Iron Steak
Garlic Bread Burgers
Grilled Flank Steak
Ground Beef Tortilla Pizza
Mongolian Lamb Barbecue
Mustard Pepper Steak
Oriental Beef Lettuce Wraps
Pepper Steak on a Stick
Peppercorn Tenderloin Steaks
Red Wine Beef Stew
Southern Comfort® Steak Marinade
Spicy Dijon Lamb Chops
Spicy Lamb Chops with Fresh Peppers and Basil
Springtime Lamb with Apricot Glaze
Stuffed Peppers
Top Sirloin Herbed Steak

Asian Beef Stir-Fry

Serves 4

This is good served over steamed brown rice with sliced water chestnuts added for extra crunch.

- 1 pound sirloin steak, thinly sliced
- 2 each red and green bell peppers, sliced ¼-inch thick
- 1 medium onion, sliced ¼-inch thick
- ½ cup dry sherry
- 2 tablespoons soy sauce, low sodium
- 1 teaspoon five-spice powder
- ¼ teaspoon black pepper
- 1 cup beef broth
- 1 tablespoon cornstarch
- 8 ounces snow pea pods, fresh
- 3 scallions, thinly sliced

1. Spray a nonstick pan with oil spray. Add meat to pan and stir-fry meat three minutes. Remove from pan.
2. Add peppers and onions to pan and stir-fry for 3 minutes. Add meat back to the pan. Add sherry and stir-fry until liquid almost evaporates, one minute.
3. Add soy sauce and beef broth to pan. Dissolve cornstarch in ½ cup water and slowly add to the pan, stirring constantly to avoid clumps. Mix in the five-spice powder and black pepper.
4. Stir sauce until it thickens enough to coat the back of a spoon.
5. Add snow peas and stir until warmed. (You want them to be crunchy.) Garnish with sliced scallions.

Nutritional Analysis
(per serving, excluding unknown items)

345	Calories
16 gm.	Fat
46.0%	Calories from Fat
27 gm.	Protein
15 gm.	Carbohydrate
3 gm.	Dietary Fiber
71 mg.	Cholesterol
687 mg.	Sodium

Exchanges

3	Lean Meat
2	Vegetable
1½	Fat

Beef and Spinach

Serves 4

1½ pounds sirloin steak, trimmed
2 teaspoons garlic powder
1 teaspoon sea salt
1 teaspoon black pepper
2 tablespoons olive oil
10 ounces spinach leaves, whole
½ teaspoon dried rosemary
1 cup Asiago cheese
Salt and pepper to taste

1. Combine garlic powder, salt, and pepper; press evenly onto beef steaks.
2. Heat 1 tablespoon olive oil in large nonstick skillet over medium heat until hot. Place steaks in skillet; cook to desired doneness, turning once. Remove; keep warm. (Steak will continue to cook.)
3. Heat remaining 1 tablespoon olive oil in same skillet over medium heat until hot. Add spinach and rosemary. Add salt and pepper to taste. Cook and stir 1–2 minutes or until spinach is just wilted. Remove from heat.
4. Add ¾ cup of the Asiago cheese to spinach; toss.
5. Serve steaks on spinach. Sprinkle with remaining ¼ cup Asiago cheese.

Nutritional Analysis
(per serving, excluding unknown items)

523	Calories
38 gm.	Fat
66.3%	Calories from Fat
40 gm.	Protein
4 gm.	Carbohydrate
2 gm.	Dietary Fiber
132 mg.	Cholesterol
943 mg.	Sodium

Exchanges

5½	Lean Meat
½	Vegetable
4½	Fat

Beef and Spinach Burritos

Serves 4

1 pound lean ground beef
1 teaspoon chili powder
½ teaspoon red chili flakes
½ teaspoon kosher salt
1¼ cups salsa, chunky
10 ounces spinach, frozen, chopped and drained
4 large flour tortillas
8 tablespoons light sour cream
2 tablespoons cilantro, chopped

1. In a large nonstick skillet, brown ground beef over medium heat for 8 to 10 minutes or until no longer pink, stirring occasionally. Drain fat.
2. Season beef with chili powder, salt, and red chili flakes. Stir in salsa and spinach. Heat thoroughly. Remove from heat.
3. To serve, spoon approximately ¼ of the beef mixture down the center of the tortilla. Fold sides to center, overlapping edges, and put seam side down on a serving plate. Spoon 2 tablespoons of sour cream on the center of the tortilla. Sprinkle with cilantro and serve.

Nutritional Analysis
(per serving, excluding unknown items)

588	Calories
30 gm.	Fat
45.5%	Calories from Fat
30 gm.	Protein
50 gm.	Carbohydrate
6 gm.	Dietary Fiber
87 mg.	Cholesterol
1077 mg.	Sodium

Exchanges

2½	Grain (Starch)
3	Lean Meat
1½	Vegetable
4	Fat

Curried Lamb Chops

Serves 4

2 pounds lamb chop, lean, 1-inch thick
⅓ cup lemon juice
¼ cup finely chopped onion
1 tablespoon olive oil
½ teaspoon curry powder
¼ teaspoon lemon peel
¼ teaspoon black pepper
¼ teaspoon ground cumin
¼ teaspoon ground cardamom
¼ teaspoon ground ginger

1. Arrange lamb chops in glass baking dish. In bowl, combine lemon juice, onion, oil, curry powder, lemon peel, black pepper, cumin, cardamom, and ginger; mix well. Pour marinade over lamb; cover and refrigerate. Marinate 6 to 24 hours; turn occasionally.
2. Broil 4 inches from heat source, 5 minutes on each side or to desired temperature.

Nutritional Analysis
(per serving,
excluding unknown items)

597	Calories
51 gm.	Fat
77.9%	Calories from Fat
30 gm.	Protein
3 gm.	Carbohydrate
Trace	Dietary Fiber
133 mg.	Cholesterol
101 mg.	Sodium

Exchanges

4	Lean Meat
8	Fat

Flat Iron Steak

Serves 4

1½ pounds sirloin steaks, trimmed, 4 steaks flat iron
2 tablespoons olive oil
2 tablespoons lime juice
1 teaspoon garlic paste, in a tube
½ teaspoon kosher salt
½ teaspoon black pepper

1. Combine oil, lime juice, and garlic paste.
2. Season beef steaks with salt and pepper.
3. Pour marinade over beef and turn with tongs to cover.
4. Refrigerate 20 minutes to 2 hours.
5. Cook on the indoor or outdoor grill to desired temperature. Let rest 5 minutes and serve.

Nutritional Analysis
(per serving,
excluding unknown items)

407	Calories
30 gm.	Fat
67.8%	Calories from Fat
31 gm.	Protein
1 gm.	Carbohydrate
Trace	Dietary Fiber
107 mg.	Cholesterol
322 mg.	Sodium

Exchanges

4½	Lean Meat
3½	Fat

Garlic Bread Burgers

Serves 4

1 loaf Italian bread

Olive oil spray

1 teaspoon garlic salt

1 pound lean ground beef

½ teaspoon kosher salt

½ teaspoon black pepper

1 whole tomato

12 whole basil leaves

4 ounces fat-free mozzarella cheese, slices

1. Cut the Italian bread loaf in half lengthwise, spray with olive oil, and sprinkle with garlic salt.
2. Form ground beef in oblong patties, sprinkle with salt and pepper. Cook on the grill until the temperature reaches 160°F.
3. Place Italian bread on the grill with sprayed side down, and grill until nicely toasted. Remove from grill, place the grilled sides together and cut in four pieces.
4. Place burgers in buns; top with mozzarella cheese, tomato, and basil. Serve.

Nutritional Analysis (per serving, excluding unknown items)

615	Calories
28 gm.	Fat
41.0%	Calories from Fat
30 gm.	Protein
59 gm.	Carbohydrate
4 gm.	Dietary Fiber
85 mg.	Cholesterol
1491 mg.	Sodium

Exchanges

4	Grain (Starch)
3	Lean Meat
½	Vegetable
3½	Fat

Grilled Flank Steak

Serves 4

Flank steak has great flavor and is best cooked medium rare.

1 pound flank steak
1 teaspoon olive oil
½ teaspoon sea salt
1 tablespoon black pepper

1. Coat each side of flank steak with oil, salt, and pepper.
2. Cook on a hot grill, 4 to 5 minutes on each side. Let rest for 10 minutes.
3. Slice on the angle.

Nutritional Analysis (per serving, excluding unknown items)

214	Calories
13 gm.	Fat
55.8%	Calories from Fat
22 gm.	Protein
1 gm.	Carbohydrate
Trace	Dietary Fiber
58 mg.	Cholesterol
315 mg.	Sodium

Exchanges

3	Lean Meat
1	Fat

Ground Beef Tortilla Pizza

Serves 4

1 pound lean ground beef
1 medium onion, chopped
1 teaspoon dried oregano
4 each flour tortillas
1 medium tomato, chopped
1 tablespoon fresh basil leaves, sliced ½-inch thick
½ cup shredded nonfat mozzarella cheese
¼ cup Parmesan cheese

1. Preheat oven to 400°F.
2. Brown ground beef and onion in a skillet over medium heat (approximately 8 to 10 minutes). Drain ground beef and blot with a paper towel. Stir oregano into beef.
3. Spray tortillas with olive oil spray and bake on large baking sheet for three minutes.
4. Spoon beef mixture evenly over top of each tortilla; top with tomato.
5. Sprinkle with basil and cheese. Return to oven and bake 12–14 minutes or until tortillas are lightly browned. Serve immediately.

Nutritional Analysis
(per serving, excluding unknown items)

595	Calories
30 gm.	Fat
46.5%	Calories from Fat
34 gm.	Protein
45 gm.	Carbohydrate
3 gm.	Dietary Fiber
92 mg.	Cholesterol
631 mg.	Sodium

Exchanges

2½	Grain (Starch)
4	Lean Meat
½	Vegetable
4	Fat

Mongolian Lamb Barbecue

Serves 6

- 3 pounds lamb chop, lean, cut 1-inch thick
- ¾ teaspoon salt
- ½ teaspoon pepper
- ¼ cup olive oil
- 3 tablespoons lemon juice
- 3 tablespoons soy sauce
- 2 tablespoons hoisin sauce
- 1 tablespoon sesame seeds, toasted
- 1 teaspoon grated ginger root

1. Salt and pepper lamb. In a 13x9-inch dish, mix together oil, juice, soy sauce, hoisin sauce, sesame seeds, and ginger root. Pour marinade over the lamb chops and coat both sides. Cover and refrigerate 4 to 24 hours. Turn occasionally. Remove lamb from marinade and discard marinade.
2. To broil: Cook lamb 4 inches from heat source. Broil for 6 to 8 minutes per side or to desired temperature.
3. To grill: Cook lamb on medium-hot grill for 6 to 8 minutes per side or to desired temperature.

Nutritional Analysis (per serving, excluding unknown items)

663	Calories
58 gm.	Fat
79.0%	Calories from Fat
30 gm.	Protein
4 gm.	Carbohydrate
Trace	Dietary Fiber
133 mg.	Cholesterol
968 mg.	Sodium

Exchanges

4	Lean Meat
9	Fat

Mustard Pepper Steak

Serves 4

SAUCE:

¼ cup apple juice

2 tablespoons parsley

2 tablespoons Dijon mustard

1 teaspoon garlic paste

1 teaspoon black peppercorns, coarsely ground

24 ounces sirloin steaks, trimmed

1. Combine sauce ingredients in small bowl. Remove and reserve ¼ cup for basting.
2. Brush steaks with remaining sauce.
3. Place steaks on grill over medium heat. Grill 12 to 18 minutes for medium rare to medium, turning occasionally. Baste steaks with reserved ¼ cup sauce during last 10 minutes of grilling.
4. Remove steaks from grill and serve.

Nutritional Analysis
(per serving,
excluding unknown items)

362	Calories
24 gm.	Fat
60.3%	Calories from Fat
32 gm.	Protein
3 gm.	Carbohydrate
1 gm.	Dietary Fiber
107 mg.	Cholesterol
183 mg.	Sodium

Exchanges

4½	Lean Meat
2	Fat

Oriental Beef Lettuce Wraps

Serves 4

1 pound ground lean beef
½ cup hoisin sauce
½ cup peanut sauce
1 medium cucumber, whole, seeded and chopped
¼ cup mint leaves, chopped
½ teaspoon sea salt
½ teaspoon red chili flakes
¼ teaspoon black pepper
12 large Boston lettuce leaves

1. Brown ground beef in large nonstick skillet over medium heat, 8 to 10 minutes, breaking up into small crumbles. Drain fat. Stir in hoisin sauce and peanut sauce; heat thoroughly.
2. Just before serving, add cucumber and chopped mint leaves; toss gently.
3. Season with salt and pepper. Serve beef mixture in lettuce leaves.

Nutritional Analysis (per serving, excluding unknown items)

502	Calories
34 gm.	Fat
61.2%	Calories from Fat
27 gm.	Protein
22 gm.	Carbohydrate
3 gm.	Dietary Fiber
86 mg.	Cholesterol
927 mg.	Sodium

Exchanges

3½	Lean Meat
½	Vegetable
4½	Fat
1½	Other Carbohydrates

Pepper Steak on a Stick

Serves 6

2 pounds skirt steak
½ cup red wine vinegar
½ cup apple juice
1 tablespoon onion flakes
1 tablespoon dried sage
1 tablespoon black pepper
1 tablespoon coriander
1 tablespoon dry mustard
1 teaspoon kosher salt
½ cup olive oil
2 whole red bell peppers, seeded and quartered
1 whole green bell pepper, seeded and quartered
6 skewers (if you are using wooden skewers, soak in water 15 minutes before using)

1. Place steak in a glass bowl. In another bowl, combine wine vinegar, apple juice, onion flakes, sage, pepper, coriander, dry mustard, salt, and olive oil. Reserve ½ cup marinade to brush on steak while grilling. Pour remaining marinade over steak, turning to coat both sides. Cover, place in refrigerator and marinate at least one hour.
2. Preheat grill to medium temperature.
3. Remove steak from marinade, discarding marinade. Cut steak into 6 strips, cutting against the grain. Thread meat onto long skewers, weaving meat around quartered peppers.
4. Grill for 12 to 15 minutes, on either an indoor or outdoor grill, turning occasionally to cook all sides. Brush meat with reserved marinade as it cooks.

Nutritional Analysis (per serving, excluding unknown items)

454	Calories
34 gm.	Fat
67.8%	Calories from Fat
30 gm.	Protein
7 gm.	Carbohydrate
1 gm.	Dietary Fiber
77 mg.	Cholesterol
421 mg.	Sodium

Exchanges

4½	Lean Meat
4½	Fat

Peppercorn Tenderloin Steaks

Serves 4

1 pound beef tenderloin, 4 steaks
½ teaspoon black peppercorns, cracked
1 tablespoon cornstarch
1 tablespoon cold water
14 ounces beef broth
⅛ teaspoon dried thyme
1 whole bay leaf
⅓ cup red wine
¼ teaspoon black peppercorns, crushed

1. Press cracked peppercorns into steaks. Cook on grill or under boiler until done.
2. Put cornstarch into a cup. Add water, stirring to dissolve.
3. In a medium saucepan, whisk together cornstarch mixture and beef broth.
4. Bring to a boil over high heat; cook until slightly thickened, stirring occasionally. Stir in bay leaf and thyme. Reduce heat to medium-high and cook for 10 to 12 minutes, or until sauce is reduced to approximately 1 cup.
5. Stir in red wine and ¼ teaspoons pepper corns. Reduce heat and simmer for 5 minutes. Remove bay leaf.
6. Place steak either on one serving plate, or four individual plates. Spoon sauce over top of steak and serve.

Nutritional Analysis
(per serving, excluding unknown items)

370	Calories
26 gm.	Fat
67.0%	Calories from Fat
25 gm.	Protein
4 gm.	Carbohydrate
Trace	Dietary Fiber
81 mg.	Cholesterol
584 mg.	Sodium

Exchanges

3	Lean Meat
3½	Fat

Red Wine Beef Stew

Serves 4

This recipe can also be cooked all day on low in a slow cooker.

1 pound stew meat
½ teaspoon kosher salt
½ teaspoon black pepper
1 tablespoon flour
1 tablespoon grapeseed oil
4 medium carrots, sliced 1-inch thick
2 medium onions, sliced ½-inch thick
1 cup red wine
15 ounces canned tomatoes
2 each rosemary sprigs
2 medium Yukon Gold potatoes, quartered

1. Preheat over to 325°F.
2. Place stew meat in a bag with kosher salt, pepper, and flour. Shake to coat meat.
3. Heat oil in a Dutch oven and brown meat. Pour in red wine and stir to get the brown bits from bottom of pan. Add remaining ingredients.
4. Cover and bake for 2 hours in oven.

Nutritional Analysis (per serving, excluding unknown items)

539	Calories
27 gm.	Fat
49.4%	Calories from Fat
35 gm.	Protein
28 gm.	Carbohydrate
6 gm.	Dietary Fiber
113 mg.	Cholesterol
794 mg.	Sodium

Exchanges

4½	Lean Meat
3	Vegetable
2½	Fat

Southern Comfort® Steak Marinade

___
Serves 8
___

½ cup Southern Comfort®
¼ cup red wine
1 teaspoon garlic powder
½ teaspoon cayenne pepper
½ teaspoon black pepper
¼ cup olive oil
8 sirloin steaks (6 oz), trimmed

1. Whisk marinade ingredients together in medium bowl. Pour ¾ of marinade over steaks in shallow container, cover, and refrigerate 1 hour.
2. Preheat grill to medium-high heat.
3. Grill steaks as desired, basting occasionally with small reserved portion of marinade.

Nutritional Analysis
(per serving,
excluding unknown items)

325	Calories
21 gm.	Fat
68.8%	Calories from Fat
19 gm.	Protein
2 gm.	Carbohydrate
Trace	Dietary Fiber
65 mg.	Cholesterol
58 mg.	Sodium

Exchanges

2½	Lean Meat
2½	Fat

Spicy Dijon Lamb Chops

Serves 4

2 pounds lamb chops, lean, sliced 1-inch thick

MARINADE:
2 tablespoons olive oil
2½ tablespoons Dijon-style mustard
2 tablespoons lemon juice
1 tablespoon worcestershire sauce
1 tablespoon finely chopped garlic
1½ teaspoons red pepper sauce
½ teaspoon paprika
½ teaspoon salt

1. In a 13x9-inch baking dish, combine all marinade ingredients.
2. Place the lamb chops in dish and coat well with the marinade. Cover and refrigerate for 6 to 8 hours, turning occasionally.
3. To broil: Cook lamb 4 inches from heat source. Broil for 5 to 6 minutes per side or to desired temperature.
4. To grill: Cook lamb over medium-hot coals. Grill 4 inches from coals for 5 to 6 minutes on each side or to desired temperature.
5. Cover and let stand for 10 minutes. Internal temperature will rise approximately 10°F.

Nutritional Analysis
(per serving,
excluding unknown items)

332	Calories
18 gm.	Fat
49.5%	Calories from Fat
38 gm.	Protein
3 gm.	Carbohydrate
Trace	Dietary Fiber
118 mg.	Cholesterol
589 mg.	Sodium

Exchanges

5½	Lean Meat
1½	Fat

Spicy Lamb Chops with Fresh Peppers and Basil

Serves 4

- 2 pounds lamb chops, lean, 1-inch thick
- 1 teaspoon lemon peel
- 1 teaspoon black pepper
- 2 teaspoons garlic powder
- 2 teaspoons dried rosemary leaves, crushed
- 2 tablespoons olive oil, divided
- 1 small red bell pepper, thinly sliced
- ½ cup chopped onion
- 3 cloves garlic, finely chopped
- ¼ cup slivered almonds, toasted and chopped
- ¼ cup chopped fresh basil
- 1 tablespoon red wine vinegar
- ½ teaspoon seasoned salt
- ½ teaspoon seasoned pepper

1. Season lamb chops with lemon peel, black pepper, garlic powder, and rosemary.
2. In large skillet, heat 1½ tablespoons oil and sauté chops, browning well on each side and cooking to desired temperature.
3. Remove chops from skillet to heat-proof platter; cover and keep warm.
4. Wipe out skillet and heat remaining oil. Add bell pepper, onion, and garlic.
5. Quickly stir-fry for 3 to 4 minutes.
6. Add almonds, basil, vinegar, salt, and pepper and stir-fry an additional 2 to 3 minutes.
7. Pour over chops and serve.

---

*Nutritional Analysis
(per serving,
excluding unknown items)*

699	Calories
59 gm.	Fat
76.7%	Calories from Fat
32 gm.	Protein
8 gm.	Carbohydrate
2 gm.	Dietary Fiber
133 mg.	Cholesterol
275 mg.	Sodium

Exchanges

4½	Lean Meat
1	Vegetable
9½	Fat

Springtime Lamb with Apricot Glaze

Serves 4

½ cup apricot jam
1 tablespoon prepared ground horseradish
1 tablespoon prepared mustard
1½ pounds lamb sirloin
1 teaspoon olive oil
2 teaspoons chopped fresh rosemary
2 cloves garlic, finely chopped
1 teaspoon seasoned pepper

1. Preheat oven to 350°F.
2. In a small bowl, blend apricot jam, horseradish, and mustard; set aside.
3. Using a roasting pan with rack, place lamb on rack. Brush with olive oil and rub with rosemary, and seasoned pepper.
4. Roast lamb in the oven for 35 to 45 minutes or to desired degree of doneness: 145°F for medium-rare, 160°F for medium, or 170°F for well. After 15 minutes of roasting, brush with apricot glaze several times.
5. Cover and let stand for 10 minutes. Internal temperature will rise approximately 10°F.

Nutritional Analysis
(per serving, excluding unknown items)

427	Calories
25 gm.	Fat
52.7%	Calories from Fat
23 gm.	Protein
27 gm.	Carbohydrate
1 gm.	Dietary Fiber
94 mg.	Cholesterol
132 mg.	Sodium

Exchanges

3½	Lean Meat
3	Fat
2	Other Carbohydrates

Stuffed Peppers

Serves 6

10 ounces bratwurst, Johnsonville® Hot 'n Spicy (casings removed)
10 ounces ground turkey
1 cup bread cubes
1 cup frozen mixed vegetables, thawed
1 can tomato sauce
1½ teaspoons Italian seasoning
½ teaspoon red chili flakes
3 green bell peppers, each cut in half
1 cup mozzarella cheese, part skim milk, shredded

1. Preheat oven to 425°F.
2. In a skillet, cook and crumble bratwurst and ground turkey until brown.
3. Drain fat. Combine meat mixture with bread cubes, vegetables, tomato sauce, Italian seasoning, and red chili flakes. Stir in mozzarella cheese.
4. Fill pepper halves with meat mixture and place in a 13x9-inch baking pan.
5. Add ¼ cup water to bottom of pan, cover, and bake for 30 to 35 minutes, or until hot.

Nutritional Analysis
(per serving, excluding unknown items)

334	Calories
20 gm.	Fat
53.5%	Calories from fat
23 gm.	Protein
16 gm.	Carbohydrate
3 gm.	Dietary Fiber
76 mg.	Cholesterol
713 mg.	Sodium

Exchanges

½	Grain (Starch)
3	Lean Meat
2	Vegetable
2	Fat

Top Sirloin Herbed Steak

Serves 4

- ¼ cup A.1.® steak sauce
- 2 teaspoons fresh basil, chopped
- 2 teaspoons fresh rosemary sprigs, chopped
- ½ teaspoon black pepper
- ⅛ teaspoon garlic powder
- 1½ pounds top sirloin steak

1. Mix steak sauce, basil, rosemary, pepper, and garlic powder; brush on both sides of steak.
2. Grill or broil steak 5 to 7 minutes on each side or until desired temperature.
3. Let rest 10 minutes. (The steak will continue cooking while resting.)
4. Slice steak and serve.

Nutritional Analysis
(per serving, excluding unknown items)

222	Calories
7 gm.	Fat
30.1%	Calories from Fat
35 gm.	Protein
3 gm.	Carbohydrate
Trace	Dietary Fiber
99 mg.	Cholesterol
341 mg.	Sodium

Exchanges

5	Lean Meat

Desserts

Angle Food Cake with Mascarpone Coffee Whip
Apple Crisp
Apple Tart
Apricot Cake
Berry Tart
Brown Rice Pudding
Cherry Cheese Delight
Cherry Tiramisu
Chocolate Bundles
Chocolate Ricotta Pudding
Low-Fat Trifle
Peach Tart
Pear and Apricot Compote
Pound Cake and Pudding
Pumpkin Cups
Strawberries Romanoff
Vanilla Ricotta Pudding

Angle Food Cake with Mascarpone Coffee Whip

Serves 6

2 tablespoons boiling water
2 tablespoons instant coffee
¾ cup mascarpone cheese
2 cups Cool Whip Lite®
12 ounces angel food cake, whole, purchased
½ cup coffee liqueur
2 tablespoons cocoa powder

1. Stir the water and instant coffee in a large bowl to blend. Using a large rubber spatula, fold in mascarpone.
2. Next, gradually fold in Cool Whip Lite® until mixture is well blended.
3. Cut the angel food cake into 12 wedges. Brush 1 side of each wedge of cake with the liqueur. Arrange 2 wedges of cake (liqueur side up) on each plate. Dollop the mascarpone coffee whip on top of each wedge of cake.
4. Dust with the cocoa powder and serve.

Nutritional Analysis (per serving, excluding unknown items)

193	Calories
10 gm.	Fat
58.0%	Calories from Fat
1 gm.	Protein
14 gm.	Carbohydrate
1 gm.	Dietary Fiber
19 mg.	Cholesterol
27 mg.	Sodium

Exchanges

2	Fat
½	Other Carbohydrates

Apple Crisp

Serves 8

2 pounds sliced apples
½ cup sugar substitute
1 tablespoon lemon juice
½ cup butter, soft
½ cup all-purpose flour
2 cups rolled oats, quick-cooking, uncooked
½ cup brown sugar

1. Preheat oven to 350°F.
2. Mix sugar substitute and lemon juice with apples. Arrange in a greased baking dish.
3. Combine remaining ingredients and mix until crumbly. Spread evenly over apples. Bake for 45 to 50 minutes. Serve with whipped cream, ice cream, or cheese.

Nutritional Analysis
(per serving,
excluding unknown items)

333	Calories
13 gm.	Fat
34.9%	Calories from Fat
4 gm.	Protein
51 gm.	Carbohydrate
5 gm.	Dietary Fiber
31 mg.	Cholesterol
146 mg.	Sodium

Exchanges

1½	Grain (Starch)
1	Fruit
2½	Fat
1	Other Carbohydrates

Apple Tart

Serves 6

1 whole Pillsbury® pie crust
½ cup apple butter
8 ounces frozen apple slices, thawed
½ teaspoon cinnamon
¼ cup sugar substitute
¼ cup walnuts, chopped
1 tablespoon sugar

1. Preheat over to 350°F.
2. Lay out pie crust.
3. Spread apple butter in center of crust, leaving a 2-inch outer circle.
4. Mix apples with cinnamon and sugar substitute. Place apples over apple butter. Sprinkle with walnuts.
5. Fold outer edges of pie crust over peaches. Brush with water and sprinkle sugar over raw crust.
6. Bake at until crust is golden brown, about 25 minutes.

Nutritional Analysis
(per serving, excluding unknown items)

283	Calories
12 gm.	Fat
38.8%	Calories from Fat
3 gm.	Protein
41 gm.	Carbohydrate
1 gm.	Dietary Fiber
7 mg.	Cholesterol
151 mg.	Sodium

Exchanges

1	Grain (Starch)
½	Fruit
2½	Fat
1	Other Carbohydrates

Apricot Cake

Serves 8

15 ounces canned apricot halves, in light syrup
3 whole eggs
½ cup sour cream, light
2 teaspoons vanilla extract
2 teaspoons Bisquick® baking mix
½ cup sugar substitute
1 cup chopped walnuts

1. Preheat oven to 350°F.
2. Place apricot halves with liquid, eggs, sour cream, vanilla, and allspice into food processor and mix well.
3. Add Bisquick® and sugar, pulse until all is mixed together.
4. Add ⅔ cup of the walnuts and pulse twice just to mix them in.
5. Spray a Bundt® pan with non-stick spray. Sprinkle the remaining ⅓ cup of walnuts into the pan. Spoon the batter into the pan. Bake for 25 minutes. Cool for 10 minutes before unmolding.

Nutritional Analysis
(per serving,
excluding unknown items)

172	Calories
11 gm.	Fat
56.9%	Calories from Fat
7 gm.	Protein
12 gm.	Carbohydrate
2 gm.	Dietary Fiber
81 mg.	Cholesterol
63 mg.	Sodium

Exchanges

1	Lean Meat
1½	Fat
½	Other Carbohydrates

Berry Tart

Serves 6

1 whole Pillsbury® pie crust
½ cup blackberry jam
8 ounces berries, mixed and thawed
¼ cup sugar substitute
¼ cup pecans, sliced
1 tablespoon sugar

1. Preheat over to 350°F.
2. Lay out pie crust.
3. Spread blackberry jam in center of crust, leaving a 2-inch outer circle.
4. Mix berries with sugar substitute. Place berries over jam. Sprinkle with pecans.
5. Fold outer edges of pie crust over berries. Brush with water and sprinkle sugar over raw crust.
6. Bake at until crust is golden brown, about 25 minutes.

Nutritional Analysis (per serving, excluding unknown items)

289	Calories
13 gm.	Fat
38.3%	Calories from Fat
2 gm.	Protein
43 gm.	Carbohydrate
1 gm.	Dietary Fiber
7 mg.	Cholesterol
160 mg.	Sodium

Exchanges

1	Grain (Starch)
2½	Fat
1½	Other Carbohydrates

Brown Rice Pudding

Serves 6

¾ cup egg whites
¾ cup sugar substitute
1½ cups milk, 1% lowfat
1 teaspoon vanilla extract
½ teaspoon almond extract
2 cups brown rice, cooked
1 cup Cool Whip Lite®

1. Preheat to 325°F.
2. Mix all ingredients, except Cool Whip Lite®, together and pour into 10-inch pan sprayed with nonstick spray.
3. Place pan in a larger pan filled halfway with water. Bake for one hour.
4. Let cool. Either serve room temperature or chill. Top with Cool Whip Lite®.

Nutritional Analysis
(per serving, excluding unknown items)

191	Calories
3 gm.	Fat
12.7%	Calories from Fat
7 gm.	Protein
32 gm.	Carbohydrate
1 gm.	Dietary Fiber
2 mg.	Cholesterol
137 mg.	Sodium

Exchanges

1	Grain (Starch)
½	Lean Meat
½	Non-fat Milk
½	Fat
1	Other Carbohydrates

Cherry Cheese Delight

_____

Serves 10

_____

1 16-ounce container low-fat cottage cheese

1 21-ounce can cherry pie filling

½ cup slivered almonds

1 teaspoon almond extract

1 cup Cool Whip Lite®

Fresh mint leaves

1. Using a food processor with steel blade whip low-fat cottage cheese until smooth.
2. Combine cherry pie filling, cottage cheese, almonds and almond extract in a large mixing bowl mix well. Fold in Cool Whip Lite® topping. Let chill until ready to serve. Garnish with mint leaves, if desired. Serve as a salad, dessert or snack.
3. Serving size: ½ cup

Nutritional Analysis
(per serving,
excluding unknown items)

168	Calories
6 gm.	Fat
32.6%	Calories from Fat
7 gm.	Protein
22 gm.	Carbohydrate
1 gm.	Dietary Fiber
2 mg.	Cholesterol
192 mg.	Sodium

Exchanges

1	Lean Meat
1	Fat
1	Other Carbohydrates

Cherry Tiramisu

Serves 8

1 cup ricotta cheese, part skim milk
½ cup sugar substitute
¼ cup sour cream, light
¼ cup coffee liqueur
1½ cups ladyfinger cookies, broken up
1 21-ounce can cherry pie filling
Fresh mint leaves for garnish (optional)

1. Place ricotta cheese, sugar substitute, sour cream, and coffee liqueur in a food processor with steel blade, whip until smooth.
2. Remove six cherries from cherry pie filling; reserve for garnish.
3. To assemble dessert, spoon 2 tablespoons ricotta cheese mixture into each of 8 (6-ounce) parfait or wine glasses.
4. Add 2 tablespoons ladyfinger cookie crumbs to each glass; top each with 2 tablespoons cherry pie filling.
5. Repeat ricotta cheese mixture, ladyfinger cookie crumbs, and cherry pie filling layers. Finish each serving with an equal portion of the remaining ricotta cheese mixture.
6. Garnish with reserved cherries, and mint leaves, if desired. Let chill 2 to 3 hours before serving.

Nutritional Analysis
(per serving,
excluding unknown items)

409	Calories
13 gm.	Fat
30.3%	Calories from Fat
7 gm.	Protein
63 gm.	Carbohydrate
1 gm.	Dietary Fiber
21 mg.	Cholesterol
249 mg.	Sodium

Exchanges

½	Lean Meat
2½	Fat
4	Other Carbohydrates

Chocolate Bundles

Serves 10

1 sheet puff pastry sheet

32 ounces chocolate chips

2 each egg whites

2 tablespoons water

2 tablespoons sugar

1. Preheat oven to 350°F.
2. Cut pastry sheet into 32 squares.
3. Place 1 ounce of chocolate chips in the center of each square. Bring ends up to center to create little bundles and twist gently.
4. Line a cookie sheet with parchment paper. Mix egg whites and water together and brush on each bundle. Sprinkle with sugar.
5. Bake for 10 to 12 minutes. Let sit 5 minutes. Remove to plate and serve three bundles to each person. (Two will be left for the cook!)

Nutritional Analysis
(per serving,
excluding unknown items)

581	Calories
36 gm.	Fat
51.3%	Calories from Fat
6 gm.	Protein
71 gm.	Carbohydrate
6 gm.	Dietary Fiber
0 mg.	Cholesterol
82 mg.	Sodium

Exchanges

$1/2$	Grain (Starch)
$7 1/2$	Fat
4	Other Carbohydrates

Chocolate Ricotta Pudding

___
Serves 6
___

Quick and easy dessert or snack.

8 ounces ricotta cheese, part skim milk, whipped
2 cups milk, 1% lowfat
1½ ounces instant pudding mix, chocolate
1 tablespoon cocoa
½ teaspoon coffee powder, instant
2 teaspoons vanilla extract
1 teaspoon almond extract
2 tablespoons almonds, toasted, chopped

1. Whip pudding into ricotta cheese. Add milk, cocoa, instant coffee, vanilla, and almond extract. Beat for 2 minutes.
2. Refrigerate for 5 to 10 minutes.
3. Spoon into dessert dishes and top with toasted almonds.

Nutritional Analysis (per serving, excluding unknown items)

137	Calories
6 gm.	Fat
36.3%	Calories from Fat
8 gm.	Protein
14 gm.	Carbohydrate
1 gm.	Dietary Fiber
15 mg.	Cholesterol
191 mg.	Sodium

Exchanges

½	Lean Meat
½	Non-fat Milk
½	Fat
½	Other Carbohydrates

Low-Fat Trifle

Serves 8

2 12-ounce packages whole frozen raspberries, thawed
½ cup sugar substitute, or to taste
½ whole angel food cake, cut into 1-inch cubes
8 teaspoons sherry (cream sherry)
8 ounces Cool Whip Lite®
6 each ginger snaps, crushed

1. Process thawed raspberries and sugar substitute in food processor or blender until smooth, about 10 seconds. Set aside.
2. Cover bottom of dessert glasses with approximately 1½ tablespoons raspberry purée. Fill class half full with angle food cake cubes.
3. Drizzle with sherry. Spoon 3 to 4 tablespoons raspberry purée over cake. Fill remainder of glass with raspberries, leaving room for topping.
4. Top with Cool Whip Lite®. Sprinkle with crushed ginger snaps.

Nutritional Analysis
(per serving,
excluding unknown items)

307	Calories
4 gm.	Fat
13.1%	Calories from Fat
3 gm.	Protein
61 gm.	Carbohydrate
4 gm.	Dietary Fiber
0 mg.	Cholesterol
270 mg.	Sodium

Exchanges

1½	Fruit
1	Fat
2½	Other Carbohydrates

Peach Tart

Serves 6
---

1 whole Pillsbury® pie crust
½ cup raspberry jam
½ tablespoon almond extract
8 ounces frozen peach slices, thawed
¼ cup sugar substitute
¼ cup almonds, sliced
1 tablespoon sugar

1. Preheat over to 350°F.
2. Lay out pie crust.
3. Mix almond extract into raspberry jam and spread in center of crust, leaving a 2-inch outer circle.
4. Mix peaches with sugar substitute. Place peaches over jam. Sprinkle with almonds.
5. Fold outer edges of pie crust over peaches. Brush with water and sprinkle sugar over raw crust.
6. Bake at until crust is golden brown, about 25 minutes.

Nutritional Analysis
(per serving,
excluding unknown items)

321	Calories
13 gm.	Fat
34.6%	Calories from Fat
3 gm.	Protein
50 gm.	Carbohydrate
1 gm.	Dietary Fiber
7 mg.	Cholesterol
163 mg.	Sodium

Exchanges

1	Grain (Starch)
½	Fruit
2½	Fat
1½	Other Carbohydrates

Pear and Apricot Compote

Serves 6

2 cups port wine

2 cups water

½ teaspoon cinnamon

¼ teaspoon ground cloves

4 whole fresh pears, cored and quartered

8 dried apricot halves

½ cup sliced almonds

1. Bring port, water, cinnamon, and cloves to a boil.
2. Add pears and apricots to the port.
3. Reduce heat to low and simmer for 30 minutes, or until pears are tender.
4. Remove pears to a bowl. Bring the port back to a boil and cook until liquid is reduced to thick syrup.
6. Pour the syrup over the pears and apricots and stir in chopped almonds.
7. This can be served warm or chilled.

---

*Nutritional Analysis
(per serving,
excluding unknown items)*

271 Calories
7 gm. Fat
30.7% Calories from Fat
3 gm. Protein
31 gm. Carbohydrate
4 gm. Dietary Fiber
0 mg. Cholesterol
7 mg. Sodium

Exchanges

½ Lean Meat
1½ Fruit
1 Fat

Pound Cake and Pudding

Serves 4

4 ounces pound cake slices
3 ounces Southern Comfort®
2 cups vanilla pudding
8 ounces strawberries
1 teaspoon sugar substitute
3 ounces sliced almonds

1. Lay out four slices of pound cake on a work surface. Using a pastry brush, liberally paint the cake with Southern Comfort®. Line small dessert cups or small glass bowls with cake.
2. Top with heaping spoonfuls of prepared vanilla instant pudding, about ½ cup per dessert cup. Sprinkle sliced strawberries with sugar and toss to coat. Top dessert cups with sliced berries and sliced almonds and serve or refrigerate.

Nutritional Analysis (per serving, excluding unknown items)

468	Calories
19 gm.	Fat
41.0%	Calories from Fat
10 gm.	Protein
53 gm.	Carbohydrate
3 gm.	Dietary Fiber
71 mg.	Cholesterol
524 mg.	Sodium

Exchanges

½	Grain (Starch)
½	Lean Meat
½	Fruit
3½	Fat
3	Other Carbohydrates

Pumpkin Cups

Serves 6

¼ cup egg substitute

15 ounces canned pumpkin

¾ cup sugar substitute

½ teaspoon kosher salt

2 teaspoons pumpkin pie spice

1⅔ cups evaporated skim milk

¾ cup Cool Whip Lite®

1. Preheat oven to 350°F.
2. Mix all ingredients, except Cool Whip Lite®, together well.
3. Pour into six 6-ounce custard cups. Place custard cups in a baking dish, filling baking dish halfway with water. Bake for 40 minutes.
4. Remove from oven and cool. Serve at room temperature or chilled. Top with Cool Whip Lite®.

Nutritional Analysis
(per serving, excluding unknown items)

166	Calories
3 gm.	Fat
14.1%	Calories from Fat
7 gm.	Protein
27 gm.	Carbohydrate
2 gm.	Dietary Fiber
3 mg.	Cholesterol
316 mg.	Sodium

Exchanges

1	Vegetable
½	Non-fat Milk
½	Fat
1	Other Carbohydrates

Strawberries Romanoff

Serves 4

SAUCE

1 cup plain nonfat yogurt

½ cup sugar substitute

½ teaspoon ground cinnamon

1 teaspoon vanilla extract

2 tablespoons brandy, peach

2 12-ounce pints fresh strawberries, hulled, and cut in bite-sized pieces

GARNISH

2 tablespoons pecan pieces, lightly toasted

1. In a small bowl, whisk sauce ingredients until fully mixed.
2. Refrigerate for at least 1 hour to make the sauce slightly firm.
4. Divide berries evenly into four dessert dishes.
5. Spoon ¼ cup sauce over each serving. Top with nuts.

Nutritional Analysis
(per serving,
excluding unknown items)

145   Calories
2 gm.   Fat
16.7%   Calories from Fat
4 gm.   Protein
24 gm.   Carbohydrate
1 gm.   Dietary Fiber
1 mg.   Cholesterol
51 mg.   Sodium

Exchanges

½   Non-fat Milk
½   Fat
1   Other Carbohydrates

Vanilla Ricotta Pudding

___
Serves 6
___

Quick and easy dessert or snack.

8 ounces ricotta cheese, part skim milk, whipped
2 cups milk, 1% lowfat
1½ ounces instant pudding mix, vanilla
2 teaspoons vanilla extract
1 teaspoon almond extract
2 tablespoons almonds, toasted, chopped

1. Whip pudding into ricotta cheese. Add milk, vanilla, and almond extract. Beat for 2 minutes.
2. Refrigerate for 5 to 10 minutes.
3. Spoon into dessert dishes and top with toasted almonds.

*Nutritional Analysis
(per serving,
excluding unknown items)*

135	Calories
5 gm.	Fat
36.6%	Calories from Fat
8 gm.	Protein
14 gm.	Carbohydrate
Trace	Dietary Fiber
15 mg.	Cholesterol
191 mg.	Sodium

Exchanges

½	Lean Meat
½	Non-fat Milk
½	Fat
½	Other Carbohydrates

Index

A
Al dente, 18
Alcoholic beverages, 15–16
Almond oil
   about, 2
   Spinach Salad with Almonds, 49
   Vinaigrette Salad, 40
Almonds
   Cherry Cheese Delight, 180
   Chocolate Ricotta Pudding, 183
   Peach Tart, 185
   Pear and Apricot Compote, 186
   Pound Cake and Pudding, 187
   Spinach Salad with Almonds, 49
   Vanilla Ricotta Pudding, 190
Angel Food Cake
   Low-Fat Trifle, 184
   with Mascarpone Coffee Whip, 174
Apple Crisp, 175
Apple Tart, 176
Apples
   Apple Crisp, 175
   Apple Tart, 176
   Applesauce Breakfast Cake, 34
   Pork with Apples, 143
Applesauce Breakfast Cake, 34
Apricot and Date Risotto, 54
Apricot Cake, 177
Apricot Roasted Chicken, 116
Apricots
   Apricot and Date Risotto, 54
   Apricot Cake, 177
   Apricot Roasted Chicken, 116
   Blackened Chicken Pasta Salad, 102
   Pear and Apricot Compote, 186
   Pork with Caramelized Onion and Apricots, 144–145
   Springtime Lamb with Apricot Glaze, 169
Artichokes
   Italian Sausage Artichoke Pasta, 104
   Scallops with Orzo in Packets, 92
Artisan oils, 2. *See also* individual oils
Arugula, 47
Asiago cheese
   about, 9
   Beef and Spinach, 153
Asian
   BBQ Chicken Thighs, 117
   Beef Stir Fry, 152
   Chicken Breast, 118
   Chicken Salad, 76
Asparagus
   Asparagus Quiche, 130
   Asparagus with Walnuts, 55
   Bow Ties with Asparagus and Parmesan Cheese, 103
   Salmon with Asparagus in Packets, 89
Avocado, 24

B
Baked Cauliflower, 56
Baked Eggs, 131
Baked Halibut Fillets, 84
Baked Mostaccioli with Tomato Basil Sauce, 98
Baker, bread, 5
Baking powder shelf life, 7
Baking soda shelf life, 7
Balsamic vinegar, 50
   Blackened Chicken Pasta Salad, 102
   Fresh Mozzarella and Tomatoes, 65
   Tomato Basil with Fresh Mozzarella, 51
Barbecue
   Asian BBQ Chicken Thighs, 117
   Jack Daniel's® sauce, 22
   Mongolian Lamb Barbecue, 160
   Satay Dipping Sauce, 28
Basil
   Baked Mostaccioli with Tomato Basil Sauce, 98
   Fresh Mozzarella and Tomatoes, 65
   Garlic Bread Burgers, 157
   Ground Beef Tortilla Pizza, 159
   Pasta with Parmesan Curls and Basil, 108
   Pasta with Vodka Sauce, 109
   Spicy Lamb Chops with Fresh Peppers and Basil, 168
   Tomato Basil with Fresh Mozzarella, 51
   Tomato Prosciutto Pasta, 113
Bass
   Sautéed Sea Bass, 91
   Sea Bass and Vegetables in Packets, 95
   Sea Bass Oriental in Packets, 94
BBQ. *See also* Barbecue
Beans
   Cassoulet, 137
   Couscous with Black Beans, 63
Beef
   Asian Beef Stir Fry, 152
   Beef and Spinach, 153
   Beef and Spinach Burritos, 154
   Beef and Tomato Bow Tie Pasta, 101
   Beef Sirloin Salad with Dried Cherries, 77
   Beef Strips with Spaghetti, 99
   Beef Stroganoff, 100
   Blackened Flank Steak Salad, 78
   Burgers with Grilled Onions and Salad, 79
   Flat Iron Steak, 156
   Garlic Bread Burgers, 157
   Grilled Flank Steak, 158
   Ground Beef Tortilla Pizza, 159
   Mustard Pepper Steak, 161
   Oriental Beef Lettuce Wraps, 162
   pantry items, 15
   Pepper Steak on a Stick, 163
   Peppercorn Tenderloin Steaks, 164
   Red Wine Beef Stew, 165
   Southern Comfort® Steak Marinade, 166
   Steak, Tomato, and Pepper Salad, 79
   Top Sirloin Herbed Steak, 171
Beer, 38
Beet, Orange, and Walnut Salad, 41
Bell peppers
   Asian Beef Stir Fry, 152
   Italian Sausage and Peppers, 139
   Pepper Steak on a Stick, 163
   Red Pepper Couscous, 66
   Spicy Lamb Chops with Fresh Peppers and Basil, 168
   Steak, Tomato, and Pepper Salad, 81
   Stuffed Peppers, 170
Berry Tart, 178
Beverages, 15–16
Black beans, 63
Black olives
   Chicken Breasts in Sour Cream Sauce, 121
   Tapenade, 72
Blackened Chicken Pasta Salad, 102
Blackened Flank Steak Salad, 78

Index

Blue cheese
  Beef Sirloin Salad with Dried Cherries, 77
  Buffalo Chicken, 119
Bow Ties with Asparagus and Parmesan Cheese, 103
Brandy, 189
Bratwurst
  Kabobs, 136
  Stuffed Peppers, 170
Breads
  Applesauce Breakfast Cake, 34
  baker, 5
  Beef and Spinach Burritos, 154
  Burgers with Grilled Onions and Salad, 79
  French, 35
  French Toast, 132
  Garlic Bread Burgers, 157
  Ground Beef Tortilla Pizza, 159
  Oat Bran Muffins, 36
  puff pastry, 11
  Rolled Prosciutto Sandwich, 37
  Spinach Goat Cheese Toast, 70
  Whole Wheat Beer, 38
  Whole Wheat Cheese Quesadilla, 73
Breakfast
  Applesauce Cake, 34
  Baked Eggs, 131
  French Toast, 132
  Ham and Cheese Quiche, 133
  Omelet, 134
Broccoli
  florets, 57
  slaw, 42
Brown Rice Pudding, 179
Brown Rice with Cashews, 58
Brown Rice with Spicy Pecans, 59
Brown sugar shelf life, 8
Brown-Bagging-It-Halibut, 83
Buffalo Chicken, 119
Burgers with Grilled Onions and Salad, 79
Burritos, 154
Butter shelf life, 7

C

Caesar Salad with Roasted Walnut Oil, 43
Cakes
  Angel Food Cake with Mascarpone Coffee Whip, 174
  Applesauce Breakfast, 34
  Apricot Cake, 177
  Low-Fat Trifle, 184
  Pound Cake and Pudding, 187
Capers
  Chicken Breast with Capers, 120
  Pork Piccata, 148

Carrots
  Chicken in Red Wine, 123–124
  Chicken Noodle Soup, 122
  Couscous, Peas, and Carrots, 61
  Halibut with Vegetables in Packets, 88
  Red Wine Beef Stew, 165
  Roast Chicken, 128
  Slow Cooked Pork Stew, 149
  Thai Pasta, 112
  Zucchini and Carrots Julienne, 74
Cashews, 58
Cassoulet, 137
Cauliflower
  baked, 56
  Pasta and Grilled Veggies, 107
Cereals, 16, 23
Cheddar cheese
  Asparagus Quiche, 130
  Ham and Cheese Quiche, 133
  Olive Cheddar spread, 26
Cheese. See also individual cheeses
  Angel Food Cake with Mascarpone Coffee Whip, 174
  Asiago, 9
  Asparagus Quiche, 130
  Baked Mostaccioli with Tomato Basil Sauce, 98
  Beef and Spinach, 153
  Beef Sirloin Salad with Dried Cherries, 77
  Bow Ties with Asparagus and Parmesan Cheese, 103
  Buffalo Chicken, 119
  Cherry Cheese Delight, 180
  Cherry Tiramisu, 181
  Chocolate Ricotta Pudding, 183
  Chopped Feta, Tomato, and Lettuce Salad, 44
  Fresh Mozzarella and Tomatoes, 65
  Fresh Spinach Salad with Pecans and Gorgonzola, 46
  Garlic Bread Burgers, 157
  Ground Beef Tortilla Pizza, 159
  guacamole, 24
  Ham and Cheese Quiche, 133
  Italian Sausage and Peppers, 139
  Mascarpone, 10
  Olive Cheddar Spread, 26
  Pantry items, 15
  Parmesan, 10
  Pasta with Parmesan Curls and Basil, 108
  Ricotta, 11
  Salmon Cucumber Slices, 67
  Scallop Salad, 80
  Scallops with Orzo in Packets, 92
  shelf life, 7
  Smoked Salmon Dip, 68
  Spinach Goat Cheese Toast, 70
  Spinach Noodle Bake, 111

Stuffed Peppers, 170
Tomato Basil with Fresh Mozzarella, 51
Vanilla Ricotta Pudding, 190
Walnut Roquefort Cheese Salad, 52
Whole Wheat Beer Bread, 38
Whole Wheat Cheese Quesadilla, 73
Cherries
  Beef Sirloin Salad with Dried Cherries, 77
  Cherry Cheese Delight, 180
  Cherry Tiramisu, 181
Chestnuts, water, 69, 125
Chicken. See also Poultry
  Apricot Roasted Chicken, 116
  Asian BBQ Chicken Thighs, 117
  Asian Chicken Breast, 118
  Asian Chicken Salad, 76
  Blackened Chicken Pasta Salad, 102
  Buffalo Chicken, 119
  Chicken Breast with Capers, 120
  Chicken Breasts in Sour Cream Sauce, 121
  Chicken in Red Wine, 123–124
  Chicken Noodle Soup, 122
  Chicken Wraps, 125
  Lemon Chicken and Rice, 126
  Paella, 140–141
  pantry items, 15
  Pineapple Chicken Breast, 127
  Roast Chicken, 128
Chicken Breast with Capers, 120
Chicken Breasts in Sour Cream Sauce, 121
Chicken in Red Wine, 123–124
Chicken Noodle Soup, 122
Chicken Wraps, 125
Chocolate
  Angel Food Cake with Mascarpone Coffee Whip, 174
  Chocolate Bundles, 182
  Chocolate Ricotta Pudding, 183
Chopped Feta, Tomato, and Lettuce Salad, 44
Coffee
  Angel Food Cake with Mascarpone Coffee Whip, 174
  Cherry Tiramisu, 181
  Chocolate Ricotta Pudding, 183
Compote, 186
Condiments, 16–17
Cooking terms, 18–19
Cornmeal shelf life, 7
Cottage cheese
  Cherry Cheese Delight, 180
  Guacamole, 24
  Spinach Noodle Bake, 111
Couscous
  about, 9
  Couscous and Zucchini, 60
  Couscous with Black Beans, 63

Index

Couscous with Mint, 64
Couscous, Peas, and Carrots, 61
Couscous, Plain, and Simple, 62
Red Pepper Couscous, 66
Cranberries, 80
Cream Cheese
    Salmon Cucumber Slices, 67
    Smoked Salmon Dip, 68
Creole Sausage and Brown Rice, 138
Cucumbers
    Cucumber Salad, 45
    Oriental Beef Lettuce Wraps, 166
    Salmon Cucumber Slices, 67
    Steak, Tomato, and Pepper Salad, 81
    Tabbouleh, 50
    Thai Pasta, 112
Curried Lamb Chops, 155

D

Dairy pantry items, 15
Dates, 54
Definitions, 18–19
Deglaze, 18
Desserts
    Angel Food Cake with Mascarpone Coffee Whip, 174
    Apple Crisp, 175
    Apple Tart, 176
    Apricot Cake, 177
    Berry Tart, 178
    Brown Rice Pudding, 179
    Cherry Cheese Delight, 180
    Cherry Tiramisu, 181
    Chocolate Bundles, 182
    Chocolate Ricotta Pudding, 183
    Low-Fat Trifle, 184
    Peach Tart, 185
    Pear and Apricot Compote, 186
    Pound Cake and Pudding, 187
    Pumpkin Cups, 188
    Strawberries Romanoff, 189
    Vanilla Ricotta Pudding, 190
Dips. See also Spreads
    guacamole, 24
    Satay Dipping Sauce, 28
    Smoked Salmon Dip, 68
    Yogurt cheese, 31
Dutch oven, 18

E

Eggs
    Asparagus Quiche, 130
    Baked Eggs, 131
    French Toast, 132
    Ham and Cheese Quiche, 133
    Omelet, 134
    shelf life, 7
En papillote, 18

Entrees
    Chicken, 116–128
    Eggs, 131–134
    Fish, 84–94
    Meats, 152–171
    Pasta, 98–113
    Pork, 136–149
    Poultry, 116–128
    Salads, 76–81
    Seafood, 84–94
Equipment, kitchen, 4–5

F

Feta cheese
    Chopped Feta, Tomato, and Lettuce Salad, 44
    Scallops with Orzo in Packets, 92
Fish. See also Seafood; Shellfish
    Baked Halibut Fillets, 84
    Brown-Bagging-It-Halibut, 85
    Halibut Fillets with California Vegetables in Packets, 87
    Halibut with Vegetables in Packets, 88
    Salmon Cucumber Slices, 67
    Salmon Oriental in Packets, 90
    Salmon with Asparagus in Packets, 89
    Sautéed Sea Bass, 91
    Sea Bass and Vegetables in Packets, 95
    Sea Bass Oriental in Packets, 94
    shelf life, 7
    Smoked Salmon Dip, 68
Five spice powder, 9
Flat Iron Steak, 156
Flour, 9
    oat, 10
    pantry items, 16
    self-rising, 12
    shelf life, 6, 7
    whole wheat, 12
Food processor, 4
Fowl. See Poultry
French Bread, 35
French Toast, 132
Fresh Mozzarella and Tomatoes, 65
Fresh Spinach Salad with Pecans and Gorgonzola, 46
Frozen foods, 14–15
Fruit
    Apple Crisp, 175
    Apple Tart, 176
    Applesauce Breakfast Cake, 34
    Apricot Cake, 177
    Apricot Roasted Chicken, 116
    Beef Sirloin Salad with Dried Cherries, 77
    Berry Tart, 178
    Blackened Chicken Pasta Salad, 102

Cherry Cheese Delight, 180
Cherry Tiramisu, 181
Chicken Wraps, 125
Grilled Southern Comfort® Shrimp à l'Orange, 86
Lemon Chicken and Rice, 126
Low-Fat Trifle, 184
Peach Tart, 185
Peaches and Pork, 142
Pear and Apricot Compote, 186
Pineapple Chicken Breast, 127
Pork Chops Caribbean, 146
Pork with Apples, 143
Pork with Caramelized Onion and Apricots, 144–145
Pound Cake and Pudding, 187
Raspberry sauce, 27
Spinach Arugula Salad, 47
Spinach, Raspberry, and Walnut Salad, 48
Springtime Lamb with Apricot Glaze, 169
Strawberries Romanoff, 189
Walnut Roquefort Cheese Salad, 52
Fruits
    Beet, Orange, and Walnut Salad, 41
    pantry items, 13–14
    Scallop Salad, 80

G

Garlic Bread Burgers, 157
Ginger snaps, 184
Goat Cheese, 70
Gorgonzola cheese
    Fresh Spinach Salad with Pecans and Gorgonzola, 46
    Scallop Salad, 80
Granola, 23
Grapeseed oil
    about, 2
    Blackened Chicken Pasta Salad, 102
    Chicken Breasts in Sour Cream Sauce, 121
    Snow Peas, 69
    Thai Pasta, 112
Grater, monoplane, 4
Green olives, 26
Grilled Flank Steak, 158
Grilled Southern Comfort® Shrimp à l'Orange, 86
Grilling
    Grilled Flank Steak, 158
    indoor, 4
    Pepper Steak on a Stick, 163
    Spicy Dijon Lamb Chops, 167
    vegetables, 107
Ground Beef Tortilla Pizza, 159
Guacamole, 24

Index

H

Halibut
    Baked Halibut Fillets, 84
    Brown-Bagging-It Halibut, 85
    Halibut Fillets with California Vegetables in Packets, 87
    Halibut with Vegetables in Packets, 88
Ham and Cheese Quiche, 133
Hazelnut oil, 2
Herbs. *See also* Spices
    Baked Eggs, 131
    Baked Mostaccioli with Tomato Basil Sauce, 98
    Couscous with Mint, 64
    Fresh Mozzarella and Tomatoes, 65
    Garlic Bread Burgers, 157
    Ground Beef Tortilla Pizza, 159
    Herbs de Provence, 9
    pantry items, 13, 16–17
    Pasta with Parmesan Curls and Basil, 108
    Pasta with Vodka Sauce, 109
    shelf life, 6, 8
    Spicy Lamb Chops with Fresh Peppers and Basil, 168
    Tomato Basil with Fresh Mozzarella, 51
    Tomato Prosciutto Pasta, 113
    Top Sirloin Herbed Steak, 171
Herbs de Provence, 9
Hoisin sauce
    about, 9–10
    Asian BBQ Chicken Thighs, 117
    Asian Chicken Breast, 118
    Broccoli Slaw, 42
    Chicken Wraps, 125
    Mongolian Lamb Barbecue, 160
    Oriental Beef Lettuce Wraps, 162
    Salmon Oriental in Packets, 87
    Sea Bass Oriental in Packets, 94
Honey shelf life, 8

I

Indoor grill, 4
Italian sausage
    Italian Sausage and Peppers, 139
    Italian Sausage Artichoke Pasta, 104
    Italian Sausage Pasta Stew, 105
    pantry items, 15

J

Jack Daniel's® Sauce 22

K

Kabobs, 136
Kitchen equipment, 4–5
Kitchen organization, 3–17

L

Lamb
    Curried Lamb Chops, 155
    Mongolian Lamb Barbecue, 160
    Spicy Dijon Lamb Chops, 167
    Spicy Lamb Chops with Fresh Peppers and Basil, 168
    Springtime Lamb with Apricot Glaze, 169
Leeks, 107
Lemon Chicken and Rice, 126
Lime, 146
Linguine and Shrimp, 106
Liquor. *See also* individual liquors
    about, 15–16
    Chicken Breast with Capers, 120
    Chicken Breasts in Sour Cream Sauce, 121
    Grilled Southern Comfort® Shrimp à l'Orange, 86
    Italian Sausage Artichoke Pasta, 104
    Pasta with Vodka Sauce, 109
    Peaches and Pork, 142
    Pork Chops with Mushrooms, 147
    Pork Piccata, 148
    Pork with Caramelized Onion and Apricots, 144–145
    Pound Cake and Pudding, 187
    Southern Comfort® Steak Marinade, 166
    Strawberries Romanoff, 189
Low-Fat Trifle, 184

M

Machine, bread, 5
Main dishes
    Chicken, 116–128
    Eggs, 131–134
    Fish, 84–94
    Meats, 152–171
    Pasta, 98–113
    Pork, 136–149
    Poultry, 116–128
    Salads, 76–81, 102
    Seafood, 84–94
Mandarin oranges, 41, 47
Mandolin, 4, 19
Mascarpone cheese, 10, 174
Mayonnaise, 8, 25, 30
Meats
    Asian Beef Stir Fry, 152
    Beef and Spinach, 153
    Beef and Spinach Burritos, 154
    Beef and Tomato Bow Tie Pasta, 101
    Beef Sirloin Salad with Dried Cherries, 77
    Beef Strips with Spaghetti, 99
    Beef Stroganoff, 100
    Blackened Flank Steak Salad, 78
    Bratwurst Kabobs, 136
    Burgers with Grilled Onions and Salad, 79
    Cassoulet, 137
    Creole Sausage and Brown Rice, 138
    Curried Lamb Chops, 155
    Flat Iron Steak, 156
    Garlic Bread Burgers, 157
    Grilled Flank Steak, 158
    Ground Beef Tortilla Pizza, 159
    Ham and Cheese Quiche, 133
    Italian Sausage and Peppers, 139
    Italian Sausage Artichoke Pasta, 104
    Italian Sausage Pasta Stew, 105
    Mongolian Lamb Barbecue, 160
    Mustard Pepper Steak, 161
    Oriental Beef Lettuce Wraps, 162
    Paella, 140–141
    pantry items, 15
    Peaches and Pork, 142
    Pepper Steak on a Stick, 163
    Peppercorn Tenderloin Steaks, 164
    Pork Chops Caribbean, 146
    Pork Chops with Mushrooms, 147
    Pork Piccata, 148
    Pork with Apples, 143
    Pork with Caramelized Onion and Apricots, 144–145
    prosciutto, 11
    Red Wine Beef Stew, 165
    Rolled Prosciutto Sandwich, 37
    shelf life, 8
    Slow Cooked Pork Stew, 149
    Southern Comfort® Steak Marinade, 166
    Spicy Dijon Lamb Chops, 167
    Spicy Lamb Chops with Fresh Peppers and Basil, 168
    Springtime Lamb with Apricot Glaze, 169
    Steak, Tomato, and Pepper Salad, 81
    Tomato Prosciutto Pasta, 113
    Top Sirloin Herbed Steak, 171
Mint
    Cherry Cheese Delight, 180
    Cherry Tiramisu, 181
    Couscous with Mint, 64
Mixer, stand, 4
Molasses shelf life, 8
Mongolian Lamb Barbecue, 160
Monoplane grater, 4
Mozzarella cheese
    Asparagus Quiche, 130
    Baked Mostaccioli with Tomato Basil Sauce, 98
    Fresh Mozzarella and Tomatoes, 65
    Garlic Bread Burgers, 157
    Ground Beef Tortilla Pizza, 159
    Italian Sausage and Peppers, 139
    Stuffed Peppers, 170
    Tomato Basil with Fresh Mozzarella, 51

Muffins, 36
Mushrooms
    Beef Stroganoff, 100
    Chicken in Red Wine, 123–124
    Chicken Noodle Soup, 122
    Chicken Wraps, 125
    Pork Chops with Mushrooms, 147
    Scallops in White Wine, 93
Mustard
    Mustard Pepper Steak, 161
    Spicy Dijon Lamb Chops, 167

N

Noodles. *See* Pasta
Nuts
    Almond Oil Vinaigrette Salad, 40
    almonds, 23
    Apple Tart, 176
    Apricot and Date Risotto, 54
    Apricot Cake, 177
    Asparagus with Walnuts, 55
    Beet, Orange, and Walnut Salad, 41
    Berry Tart, 178
    Brown Rice with Cashews, 58
    Brown Rice with Spicy Pecans, 59
    Caesar Salad with Roasted Walnut Oil, 43
    Cherry Cheese Delight, 180
    Chocolate Ricotta Pudding, 181
    Fresh Spinach Salad with Pecans and Gorgonzola, 46
    oils, 2
    Peach Tart, 185
    Pear and Apricot Compote, 186
    pine, 10
    Satay Dipping Sauce, 28
    shelf life, 6, 8
    Spinach Arugula Salad, 47
    Strawberries Romanoff, 189
    Vanilla Ricotta Pudding, 190
    Walnut Oil Mayonnaise, 30

O

Oat Bran Muffins, 36
Oats
    Apple Crisp, 175
    bran muffins, 36
    flour, 10
    granola, 23
Oils, artisan, 2. *See also* individual oils
Olive Cheddar Spread, 26
Olives
    Chicken Breasts in Sour Cream Sauce, 121
    Olive Cheddar Spread, 26
    Tapenade, 72
Omelet, 134
Onions
    Burgers with Grilled Onions and Salad, 79
    Pork with Caramelized Onion and Apricots, 144–145
Oranges
    Beet, Orange, and Walnut Salad, 41
    Chicken Wraps, 125
    Grilled Southern Comfort® Shrimp à l'Orange, 86
    Spinach Arugula Salad, 47
Organization, kitchen, 3–17
Oriental Beef Lettuce Wraps, 162
Orzo, 10, 92

P

Paella, 10, 140–141
Panini maker, 4
Pantry
    ingredients, 9–17
    shelf life, 6–8
Paper, cooking in
    about, 18
    Brown-Bagging-It Halibut, 85
    Halibut Fillets with California Vegetables in Packets, 87
    Halibut with Vegetables in Packets, 88
    Salmon Oriental in Packets, 90
    Salmon with Asparagus in Packets, 89
    Scallops with Orzo in Packets, 92
    Sea Bass and Vegetables in Packets, 95
    Sea Bass Oriental in Packets, 94
Parchment paper, 18
Parmesan cheese, 10
    Bow Ties with Asparagus and Parmesan Cheese, 103
    Pasta with Parmesan Curls and Basil, 108
Pasta
    Baked Mostaccioli with Tomato Basil Sauce, 98
    Beef and Tomato Bow Tie Pasta, 101
    Beef Strips with Spaghetti, 99
    Beef Stroganoff, 100
    Blackened Chicken Pasta Salad, 102
    Bow Ties with Asparagus and Parmesan Cheese, 103
    Chicken Noodle Soup, 122
    ingredients, 3
    Italian Sausage Artichoke Pasta, 104
    Italian Sausage Pasta Stew, 105
    Linguine and Shrimp, 106
    organized kitchen, 3
    orzo, 10
    pantry items, 16
    Pasta and Grilled Veggies, 107
    Pasta with Parmesan Curls and Basil, 108
    Pasta with Vodka Sauce, 109
    Scallops with Orzo in Packets, 92
    shelf life, 8
    Shrimp and Zucchini with Fettucine, 110
    Spinach Noodle Bake, 111
    Thai Pasta, 112
    Tomato Prosciutto Pasta, 113
Peaches
    Peach Tart, 185
    Peaches and Pork, 142
Peanut Butter
    Satay Dipping Sauce, 28
    Thai Pasta, 112
Pears
    Pear and Apricot Compote, 186
    Walnut Roquefort Cheese Salad, 52
Peas
    Asian Beef Stir Fry, 152
    Couscous, Peas, and Carrots, 61
    Snow Peas, 69
Pecans
    Berry Tart, 178
    Brown Rice with Spicy Pecans, 59
    Fresh Spinach Salad with Pecans and Gorgonzola, 46
    Strawberries Romanoff, 189
Pepper
    Mustard Pepper Steak, 161
    Peppercorn Tenderloin Steaks, 164
Pepper Steak on a Stick, 163
Peppercorn Tenderloin Steaks, 164
Peppers
    Asian Beef Stir Fry, 152
    Italian Sausage and Peppers, 139
    Pepper Steak on a Stick, 163
    Red Pepper Couscous, 66
    Spicy Lamb Chops with Fresh Peppers and Basil, 168
    Steak, Tomato, and Pepper Salad, 81
    Stuffed Peppers, 170
Pies. *See also* Tarts
    Spinach Potato Pie, 71
Pignoli, 10
Pine nuts, 10
Pineapple
    Grilled Southern Comfort® Shrimp à l'Orange, 86
    Pineapple Chicken Breast, 127
Pinyons, 10
Pizza, 159
Pork
    Bratwurst Kabobs, 136
    Cassoulet, 137
    Creole Sausage and Brown Rice, 138
    Ham and Cheese Quiche, 133
    Italian Sausage and Peppers, 139
    Paella, 140–141
    pantry items, 15
    Peaches and Pork, 142
    Pork Chops Caribbean, 146
    Pork Chops with Mushrooms, 147
    Pork Piccata, 148
    Pork with Apples, 143

Pork (*Continued*)
    Pork with Caramelized Onion and Apricots, 144–145
    Slow Cooked Pork Stew, 149
Port, 11, 186
Potatoes
    Chicken in Red Wine, 123–124
    Red Wine Beef Stew, 165
    Roast Chicken, 128
    Spinach Potato Pie, 71
Poultry. *See also* Chicken
    shelf life, 8
    Stuffed Peppers, 170
Pound Cake and Pudding, 187
Prosciutto
    about, 11
    Rolled Prosciutto Sandwich, 37
    Tomato Prosciutto Pasta, 113
Prunes, 116
Puddings
    Brown Rice Pudding, 179
    Chocolate Ricotta Pudding, 183
    Pound Cake and Pudding, 187
    Vanilla Ricotta Pudding, 190
Puff pastry, 11, 182
Pumpkin Cups, 188

Q

Quesadillas, 73
Quiche
    Asparagus Quiche, 130
    Ham and Cheese Quiche, 133
    pans, 19

R

R-T-C, 19
Raspberries
    Low-Fat Trifle, 184
    Spinach, Raspberry, and Walnut Salad, 48
Raspberry sauce, 27
Ready-to-cook, 19
Red Pepper Couscous, 66
Red Wine Beef Stew, 165
Reduce, 19
Rice
    Apricot and Date Risotto, 54
    Brown Rice Pudding, 179
    Brown Rice with Cashews, 58
    Brown Rice with Spicy Pecans, 59
    Creole Sausage and Brown Rice, 138
    Lemon Chicken and Rice, 126
    Paella, 140–141
    pantry items, 16
    shelf life, 6, 8
Rice vinegar, 11
Ricotta cheese
    about, 11
    Cherry Tiramisu, 181
    Chocolate Ricotta Pudding, 183
    Vanilla Ricotta Pudding, 190
Risotto, 54
Roast Chicken, 128
Rolled Prosciutto Sandwich, 37
Roquefort cheese, 52

S

Salads
    Almond Oil Vinaigrette Salad, 40
    Asian Chicken Salad, 76
    Beef Sirloin Salad with Dried Cherries, 77
    Beet, Orange, and Walnut Salad, 41
    Blackened Chicken Pasta Salad, 102
    Blackened Flank Steak Salad, 78
    Broccoli Slaw, 42
    Burgers with Grilled Onions and Salad, 79
    Caesar Salad with Roasted Walnut Oil, 43
    Chopped Feta, Tomato, and Lettuce Salad, 44
    Cucumber Salad, 45
    Fresh Spinach Salad with Pecans and Gorgonzola, 46
    ingredients, 3
    organized kitchen, 3
    pantry items, 13
    Scallop Salad, 80
    Spinach Arugula Salad, 47
    Spinach Salad with Almonds, 49
    Spinach, Raspberry, and Walnut Salad, 48
    Steak, Tomato, and Pepper Salad, 81
    Tabbouleh, 50
    Tomato Basil with Fresh Mozzarella, 51
    Walnut Roquefort Cheese Salad, 52
Salmon
    Salmon Cucumber Slices, 67
    Salmon Oriental in Packets, 90
    Salmon with Asparagus in Packets, 89
    Smoked Salmon Dip, 68
Salsa, 24, 73
Salt shelf life, 8
Sandwich, Rolled Prosciutto, 37
Satay Dipping Sauce, 28
Sauce
    Baked Mostaccioli with Tomato Basil Sauce, 98
    Barbeque, 22
    Chicken Breasts in Sour Cream Sauce, 121
    Satay Dipping Sauce, 28
    Tomato Sauce, 29
    Yogurt Cream, 31
Sausage
    Bratwurst Kabobs, 136
    Cassoulet, 137
    Creole Sausage and Brown Rice, 138
    Italian Sausage and Peppers, 139
    Italian Sausage Artichoke Pasta, 104
    Italian Sausage Pasta Stew, 105
    Paella, 140–141
    pantry items, 15
    Stuffed Peppers, 170
Sauté, 19
Sautéed Sea Bass, 91
Scallops
    Scallop Salad, 80
    Scallops in White Wine, 93
    Scallops with Orzo in Packets, 92
Sea bass
    Sautéed Sea Bass, 91
    Sea Bass and Vegetables in Packets, 95
    Sea Bass Oriental in Packets, 94
Seafood. *See also* Fish; Shellfish
    Baked Halibut Fillets, 84
    Brown-Bagging-It-Halibut, 85
    Grilled Southern Comfort® Shrimp à l'Orange, 86
    Halibut Fillets with California Vegetables in Packets, 87
    Halibut with Vegetables in Packets, 88
    Linguine and Shrimp, 106
    Paella, 140–141
    pantry items, 15
    Salmon Cucumber Slices, 67
    Salmon Oriental in Packets, 90
    Salmon with Asparagus in Packets, 89
    Sautéed Sea Bass, 91
    Scallop Salad, 80
    Scallops in White Wine, 93
    Scallops with Orzo in Packets, 92
    Sea Bass and Vegetables in Packets, 95
    Sea Bass Oriental in Packets, 94
    Shrimp and Zucchini with Fettucine, 110
    Smoked Salmon Dip, 68
Seasonings, 16–17. *See also* Herbs; Spices
Shelf life, 6–8
Shellfish. *See also* Fish; Seafood
    about, 15
    Grilled Southern Comfort® Shrimp à l'Orange, 86
    Linguine and Shrimp, 106
    Paella, 140–141
    Shrimp and Zucchini with Fettucine, 110
Shortening shelf life, 8
Shrimp
    Grilled Southern Comfort® Shrimp à l'Orange, 86
    Linguine and Shrimp, 106
    Paella, 140–141
    pantry items, 15
    Shrimp and Zucchini with Fettucine, 110

Side dishes
  Apricot and Date Risotto, 54
  Asparagus with Walnuts, 55
  Baked Cauliflower, 56
  Broccoli Florets, 57
  Brown Rice with Cashews, 58
  Brown Rice with Spicy Pecans, 59
  Couscous and Zucchini, 60
  Couscous with Black Beans, 63
  Couscous with Mint, 64
  Couscous, Peas, and Carrots, 61
  Couscous, Plain, and Simple, 62
  Fresh Mozzarella and Tomatoes, 65
  Red Pepper Couscous, 66
  Salmon Cucumber Slices, 67
  Smoked Salmon Dip, 68
  Snow Peas, 69
  Spinach Goat Cheese Toast, 70
  Spinach Potato Pie, 71
  Tapenade, 72
  Whole Wheat Cheese
    Quesadilla, 73
  Zucchini and Carrots Julienne, 74
Simmer, 19
Slaw, Broccoli, 42
Slow Cooked Pork Stew, 149
Smoked salmon, 67, 68
Smoked Salmon Dip, 68
Snow Peas, 69, 152
Soup, 122
Southern Comfort®
  about, 86
  Peaches and Pork, 142
  Pound Cake and Pudding, 187
Southern Comfort® Steak
  Marinade, 166
  Steak Marinade, 166
Spices. See also Herbs
  Curried Lamb Chops, 155
  five spice powder, 9
  Mustard Pepper Steak, 161
  pantry items, 16–17
  shelf life, 6, 8
  Spicy Lamb Chops with Fresh Peppers
    and Basil, 168
Spicy Dijon Lamb Chops, 167
Spicy Lamb Chops with Fresh Peppers
  and Basil, 198
Spinach
  Asian Chicken Salad, 76
  Beef and Spinach, 153
  Beef and Spinach Burritos, 154
  Fresh Spinach Salad with Pecans and
    Gorgonzola, 46
  Spinach Arugula Salad, 47
  Spinach Goat Cheese Toast, 70
  Spinach Noodle Bake, 111
  Spinach Potato Pie, 71
  Spinach Salad with Almonds, 49
  Spinach, Raspberry, and Walnut
    Salad, 48

Spreads. See also Dips
  mayonnaise, 25
  Tapenade, 72
Springtime Lamb with Apricot Glaze, 169
Stand mixer, 4
Steak, Tomato, and Pepper Salad, 81
Stew
  Italian Sausage Pasta Stew, 105
  Red Wine Beef Stew, 165
  Slow Cooked Pork Stew, 149
Stir fry, 19
Strawberries
  Pound Cake and Pudding, 187
  Strawberries Romanoff, 189
Stroganoff, 100
Stuffed Peppers, 170
Sugar shelf life, 8
Sun-dried tomatoes, 104

T

Tabbouleh, 50
Tapenade, 72
Tarts. See also Pies
  Apple Tart, 176
  Berry Tart, 178
  Peach Tart, 185
Temperature, 19
Terms, 18–19
Thai Pasta, 112
Tiramisu, 181
Tomato Basil with Fresh Mozzarella, 51
Tomato Prosciutto Pasta, 113
Tomato Sauce, 29
Tomatoes
  Baked Mostaccioli with Tomato Basil
    Sauce, 98
  Beef and Tomato Bow Tie Pasta, 101
  Beef Strips with Spaghetti, 99
  Blackened Flank Steak Salad, 78
  Burgers with Grilled Onions and
    Salad, 79
  Chopped Feta, Tomato, and Lettuce
    Salad, 44
  Fresh Mozzarella and Tomatoes, 65
  Italian Sausage Artichoke Pasta, 104
  Italian Sausage Pasta Stew, 105
  Pasta with Vodka Sauce, 109
  Red Wine Beef Stew, 165
  Scallops with Orzo in Packets, 92
  Shrimp and Zucchini with
    Fettucine, 110
  Slow Cooked Pork Stew, 149
  Spinach Goat Cheese Toast, 70
  Steak, Tomato, and Pepper Salad, 81
  Tabbouleh, 50
  Tomato Basil with Fresh
    Mozzarella, 51
  Tomato Prosciutto Pasta, 113
  Tomato Sauce, 29
Top Sirloin Herbed Steak, 171

Tortillas
  Beef and Spinach Burritos, 154
  Ground Beef Tortilla Pizza, 159
  Whole Wheat Cheese Quesadilla, 73
Turkey. See also Poultry
  pantry items, 15
  Stuffed Peppers, 170

V

Vanilla Ricotta Pudding, 190
Vanilla shelf life, 8
Vegetables
  Almond Oil Vinaigrette Salad, 40
  Asian Beef Stir Fry, 152
  Asian Chicken Salad, 76
  Asparagus Quiche, 130
  Asparagus with Walnuts, 55
  Baked Cauliflower, 56
  Beef and Spinach, 153
  Beef and Spinach Burritos, 154
  Beef Sirloin Salad with Dried
    Cherries, 77
  Beef Strips with Spaghetti, 99
  Beet, Orange, and Walnut Salad, 41
  Blackened Chicken Pasta Salad, 102
  Blackened Flank Steak Salad, 78
  Bow Ties with Asparagus and
    Parmesan Cheese, 103
  Bratwurst Kabobs, 136
  Broccoli Florets, 57
  Broccoli Slaw, 42
  Burgers with Grilled Onions and
    Salad, 79
  Caesar Salad with Roasted Walnut
    Oil, 43
  Chicken in Red Wine, 123–124
  Chicken Wraps, 125
  Chopped Feta, Tomato, and Lettuce
    Salad, 44
  Couscous and Zucchini, 60
  Couscous, Peas, and Carrots, 61
  Cucumber Salad, 45
  Fresh Mozzarella and Tomatoes, 65
  Fresh Spinach Salad with Pecans and
    Gorgonzola, 46
  Halibut Fillets with California
    Vegetables in Packets, 87
  Halibut with Vegetables in Packets, 88
  Italian Sausage Artichoke Pasta, 104
  Italian Sausage Pasta Stew, 105
  Mustard Pepper Steak, 161
  Oriental Beef Lettuce Wraps, 162
  pantry items, 13, 14
  Pasta and Grilled Veggies, 107
  Pork Chops with Mushrooms, 147
  Red Pepper Couscous, 66
  Red Wine Beef Stew, 165
  Roast Chicken, 128
  Salmon Oriental in Packets, 90
  Salmon with Asparagus in Packets, 89

Vegetables (*Continued*)
  Scallop Salad, 80
  Sea Bass and Vegetables in Packets, 95
  Sea Bass Oriental in Packets, 94
  shelf life, 8
  Shrimp and Zucchini with Fettucine, 110
  Slow Cooked Pork Stew, 149
  Snow Peas, 69
  Spicy Lamb Chops with Fresh Peppers and Basil, 168
  Spinach Arugula Salad, 47
  Spinach Goat Cheese Toast, 70
  Spinach Potato Pie, 71
  Spinach Salad with Almonds, 49
  Spinach, Raspberry, and Walnut Salad, 48
  Stuffed Peppers, 170
  Tabbouleh, 50
  Thai Pasta, 112
  Tomato Basil with Fresh Mozzarella, 51
  Walnut Roquefort Cheese Salad, 52
  Zucchini and Carrots Julienne, 74
Vermouth
  Chicken Breast with Capers, 120
  Chicken Breasts in Sour Cream Sauce, 121
  Italian Sausage Artichoke Pasta, 104
  Pork Chops with Mushrooms, 147
  Pork Piccata, 148

Pork with Caramelized Onion and Apricots, 144–145
Vinaigrette, 40
Vinegar
  rice, 11
  shelf life, 8
Vodka Sauce, 109

W

Walnut oil
  about, 2
  Asparagus with Walnuts, 55
  Beet, Orange, and Walnut Salad, 41
  Caesar Salad with Roasted Walnut Oil, 43
  Spinach Arugula Salad, 47
  Spinach, Raspberry, and Walnut Salad, 48
  Walnut Oil Mayonnaise, 30
Walnuts
  Apple Tart, 176
  Apricot and Date Risotto, 54
  Apricot Cake, 177
  Asparagus with Walnuts, 55
  Spinach, Raspberry, and Walnut Salad, 48
  Walnut Roquefort Cheese Salad, 52
Water chestnuts, 69, 125
Whiskey Barbeque sauce, 22
Whole Wheat Beer Bread, 38
Whole Wheat Cheese Quesadilla, 73

Whole wheat flour, 12
Wine
  about, 15–16
  Chicken in Red Wine, 123–124
  Pear and Apricot Compote, 186
  Red Wine Beef Stew, 165
  Scallops in White Wine, 90
  Southern Comfort® Steak Marinade, 166

Y

Yeast shelf life, 8
Yogurt
  Cheese, 31
  Cream, 31
  Strawberries Romanoff, 189

Z

Zesting, 4
Zucchini
  Beef Strips with Spaghetti, 99
  Couscous and Zucchini, 60
  Halibut with Vegetables in Packets, 88
  Italian Sausage Pasta Stew, 105
  Pasta and Grilled Veggies, 107
  Shrimp and Zucchini with Fettucine, 110
  Zucchini and Carrots Julienne, 74